V Columbian Conference of Engineering Physics (V CNIF)

Journal of Physics: Conference Series Volume 850

Medellin, Columbia
26-30 September 2016

ISBN: 978-1-5108-4267-0
ISSN: 1742-6588

Printed from e-media with permission by:

Curran Associates, Inc.
57 Morehouse Lane
Red Hook, NY 12571

Some format issues inherent in the e-media version may also appear in this print version.

Printed by Curran Associates, Inc. (2017)

For permission requests, please contact the Institute of Physics
at the address below.

Institute of Physics
Dirac House, Temple Back
Bristol BS1 6BE UK

Phone: 44 1 17 929 7481
Fax: 44 1 17 920 0979

techtracking@iop.org

Additional copies of this publication are available from:

Curran Associates, Inc.
57 Morehouse Lane
Red Hook, NY 12571 USA
Phone: 845-758-0400
Fax: 845-758-2633
Email: curran@proceedings.com
Web: www.proceedings.com

TABLE OF CONTENTS

IOP Conf. Series: Journal of Physics: Conf. Series **850** (2017) 011001 doi:10.1088/1742-6596/850/1/011001

Dear Readers,

The 5th Colombian Conference of Engineering Physics (V CNIF) (www.ingenieriafisica.co) was hosted on 26-30 September 2016 at Medellin, Colombia organized by *EAFIT University,* National University of Colombia and Colombian Society of Engineering Physics.

This was the Fifth version of biennial meetings that began in 2010. The five-day scientific program of the V CNIF consisted of 14 Magisterial Conferences, and 8 Courses with the participation of undergraduate and graduate students, professors, researchers and entrepreneurs from Colombia, Canada, Spain, Switzerland, United Kingdom, United States, among others. The aim of the organizing committee is that the V CNIF would provided enough opportunities for direct scientific contacts, interesting discussions and interactive exchange of ideas between the participants, for future networks.

We have had about 350 participants, with more than 300 oral and poster presentations in Breaking Research of Advanced Materials Science and Technologies, Nanomaterials, Optical, Electronic and Magnetic Materials, Multifunctional Composites and Other Advanced Materials, Graphite and Graphene, Automation and Equipment Manufacturing Technologies, Advanced Materials for Engineering Applications, Energy, Biomaterials and Healthcare from experts, academicians, senior scientists, industry executives and project leaders from 10+ countries. It is intended to provide a platform for professionals around the world to exchange state-of-the-art research and development and identify research needs and opportunities in this emerging field. As in previous versions of this event, the participants were involved in the competition for the best poster.

IOP Conf. Series: Journal of Physics: Conf. Series **850** (2017) 011001 doi:10.1088/1742-6596/850/1/011001

Certificates and cash prizes were awarded to a number of participants for the best posters selected by the Program Committee.

On behalf of the organizing and scientific committees, we are glad to thank again all the people involved in this event, who made the VCNIF a great success.

Please make a note that the **6th International Conference of Engineering Physics (VI CNIF) 2018** will take place from 24th to 28th September in the Caribbean region coast of Colombia, hosted by University of Atlántico and North University in the city of Barranquilla, Colombia.

With kind regards

The Editor

Jaime Andrés Pérez-Taborda
President of Colombian Society of Engineering Physics
sociedadcolombianaingenieriafisica@utp.edu.co

5th Colombian Conference on Engineering Physics - VCNIF

Colombian Society of Engineering Physics

EAFIT University

National University of Colombia

September 26th to 30th, 2016 Medellín – Colombia

UNIVERSIDAD EAFIT

SCIF
Sociedad Colombiana de Ingeniería Física

UNIVERSIDAD NACIONAL DE COLOMBIA

ORGANIZING INSTITUTIONS

National University of Colombia, Medellin
EAFIT University
Colombian Society of Engineering Physics

ORGANIZING COMMITTEE

ALCIDES MONTOYA CAÑOLA – Chairman V - CNIF
Head of School of Physics - National University of Colombia, Medellin

JAIME ANDRÉS PÉREZ TABORDA – Editor
President - Colombian Society of Engineering Physics

JUAN FERNANDO BOTERO-CADAVID
Curriculum Coordinator - National University of Colombia, Medellin

LILIANA PATRICIA VERA LONDOÑO
Vice-president - Colombian Society of Engineering Physics

LUIS FERNANDO RENDÓN CORTÉS
Head EAFIT Continuing Education Center

MAURICIO ARROYAVE FRANCO - Chairman V - CNIF
Head of the Department of Physical Sciences

ROGELIO OSPINA OSPINA
Regional Representative, Santander - Colombian Society of Engineering Physics

IOP Conf. Series: Journal of Physics: Conf. Series **850** (2017) 011002 doi:10.1088/1742-6596/850/1/011002

COMMITTEE SCIENTIFIC

Alcides Montoya Cañola	Escuela de Ciencias - Universidad Nacional de Colombia Sede Medellín
Álvaro Andrés Velásquez	Escuela de Ciencias – EAFIT, Colombia
Beatriz Cruz Muñoz	Departamento de Física - Universidad Tecnológica de Pereira, Colombia
Darío Alberto Castro Castro	Departamento de Física – UNINORTE, Colombia
Diego Alberto Bravo Montenegro	Departamento de Física - Universidad del Cauca, Colombia
Efraín Solarte	Departamento de Física - Universidad del Valle, Colombia
Elisabeth Restrepo Parra	Departamento de Física y Química - Universidad Nacional de Colombia, Colombia
Elvis O. López	Experimental and Applied Physics – CBPF, Brazil
Henry Riascos Landázuri	Departamento de Física - Universidad Tecnológica de Pereira, Colombia
Hernán Olaya Dávila	Facultad de ciencias - Universidad Pedagógica Y Tecnológica De Colombia
Jaime Andrés Perez Taborda	Instituto de Microelectrónica de Madrid-CSIC, Spain
Javier Salvador Rojas Montes	Facultad de Ingeniería - Universidad Santiago de Cali, Colombia
JORGE HERNAN QUINTERO	Departamento de Ciencias Básicas - Universidad de Medellín, Colombia
Jorge Iván García Sucerquia	Escuela de Ciencias - Universidad Nacional de Colombia Sede Medellín
Liliana Vera	Instituto de Microelectrónica de Madrid-CSIC, Spain
Mario Elkin Vélez Ruiz	Departamento de Ciencias Físicas - Universidad EAFIT, Colombia
Marlon Rincón Fulla	Escuela de Ciencias - Universidad Nacional de Colombia Sede Medellín
Martha Lucia Molina Prado	Facultad de Ciencias Básicas - Universidad de Pamplona, Colombia
Mauricio Arroyave Franco	Departamento de Ciencias Físicas - Universidad EAFIT, Colombia
Olga Lucía Quintero Montoya	Escuela de Ciencias – EAFIT, Colombia
OSCAR JAVIER SUAREZ SIERRA	Cinvestav - Ipn Unidad Guadalajara, México
René Restrepo	Escuela de Ciencias – EAFIT, Colombia
Román Eduardo Castañeda Sepúlveda	Escuela de Ciencias - Universidad Nacional de Colombia Sede Medellín
Sonia Gaona	Departamento de Física - Universidad del Cauca, Colombia

IOP Conf. Series: Journal of Physics: Conf. Series **850** (2017) 011002 doi:10.1088/1742-6596/850/1/011002

The Plenary Speakers of V National Conference on Engineering Physics (Alphabetically) were:

- **ANDREA FERRARI, PhD** - University of Cambridge, UK
  Graphene and related materials: from basic science to applications

- **ALESSANDRO CURIONI, PhD** - IBM Research – Zurich Lab
  Applying Physics to Industrial Challenges in the 21st Century

- **JEAN PAUL ALLAIN, PhD** - University of Illinois, Urbana-Champaign, USA
  Directed irradiation synthesis: manipulating matter in nanoscale self-organized systems

- **JOSEPH A. GIAIME, PhD** - LIGO Livingston Observatorio, USA
  Observation of gravitational waves from black hole coalescence
  Minicourse: *LIGO: Astrophysics from precision engineering*

- **MAURICIO TERRONES, PhD** - Pennsylvania State University, USA
  Nanoscience and Technology of layered materials; from doping and controlled synthesis to sensors and medical applications

- **ROBERTO MORANDOTTI, PhD** - Institut national de la recherche scientifique, Canadá
  Complex Quantum states on a chip

IOP Conf. Series: Journal of Physics: Conf. Series **850** (2017) 011002 doi:10.1088/1742-6596/850/1/011002

UNIVERSIDAD EAFIT · SCIF · UNIVERSIDAD NACIONAL DE COLOMBIA

Lunes 26 de septiembre
Universidad Nacional - sede Medellín

HORA	ACTIVIDAD	LUGAR	DESCRIPCIÓN
8:00 - 9:00	Inscripciones	Hall de la biblioteca Efe Gómez	Entrega de escarapelas y material del congreso
9:00 - 10:00*	Minicurso 1	Aula Máxima (11-202)	*Simulación Monte Carlo aplicado al estudio de materiales* // Johans Restrepo
	Minicurso 2	Auditorio Samuel Melguizo (24-307)	*Aerospace electronic instrumentation and systems design applied to aircraft avionics* // Mario Andrés Córdoba Gonzalez
	Minicurso 3	Aula 41-102	*Energía solar fotovoltaica, dimensionamiento de un sistema autosostenible* // José Ignacio Marulanda Bernal
	Minicurso 4	Aula 21-402/ 21-403	*Uso de los dispositivos móviles en los laboratorios de física básica* // Diego Aristizábal
	Minicurso 5	Aula 21-228 (sala 1)	*Fundamentos de simulación Monte Carlo para el estudio de propiedades magnéticas de materiales* // Elisabeth Restrepo Parra
	Minicurso 6	Aula TIC BL 16 109-3	*Principios de microscopia holográfica digital con y sin lentes* // Jorge Iván García Sucerquia
10:00 - 10:30	Refrigerio		Cafetería El carambolo (BL 44 - Agora)
10:30 - 12:00	Minicursos 1-6		Como se indicó en la franja*
12:00 - 14:00	Almuerzo libre		
14:00 - 15:00	Conferencia plenaria	Auditorio Samuel Melguizo (24-307) // Aula Máxima (11-202)***	*Graphene and related materials: from basic science to applications* // Andrea Ferrari
15:00 - 16:00**	Minicurso 7	Aula 21-402/ 21-403	*Uso de los dispositivos móviles en los laboratorios de física básica* // Diego Aristizábal
	Minicurso 8	Aula 12-204	*LIGO: Astrophysics from precision engineering* // Joseph A. Giaime
16:00- 16:30	Refrigerio		Cafetería El carambolo (BL 44 - Agora)
16:30 - 18:00	Minicursos 7-8		Como se indicó en la franja**
18:00 - 20:00	Coctel	Agora	Coctel de bienvenida y presentación de grupo artístico en homenaje a Pink Floyd

***Conferencia plenaria retransmitida vía streaming.

Andrea Ferrari
Director of the Cambridge Graphene Centre - Cambridge University (UK)

Martes 27 de septiembre
Universidad EAFIT

HORA	ACTIVIDAD	LUGAR	DESCRIPCIÓN
7:30 - 8:30	Registro	Auditorio Fundadores	Registro de inscritos y matriculados
8:30 - 9:00	Acto de apertura	Auditorio Fundadores	
9:00 - 10:00	Conferencia inaugural	Auditorio Fundadores	*Observation of gravitational waves from black hole coalescence* // Joseph A. Giaime
10:00 - 10:30	Refrigerio	Hall del Auditorio Fundadores // Plazoleta del Auditorio Fundadores	
10:30 - 11:30	Conferencia plenaria	Auditorio Fundadores	*Directed irradiation synthesis: manipulating matter in nanoscale self-organized systems* // Jean Paul Allain
11:30 - 12:00	Conferencia plenaria	Plazoleta del Auditorio Fundadores	Toma de fotografía del Congreso
12:00 - 14:00	Almuerzo libre		
14:00 - 14:30	Conferencia paralela	Auditorio Fundadores	*Espectroscopia óptica de plasmas producida por láser* // Henry Riascos
		Aula 27-101	*Técnicas de Modelamiento y simulación de propiedades físicas de Materiales* // Elisabeth Restrepo
		Aula 19-522	*Ultra-low thermal conductivities in large-area SiGe nanomeshes for thermoelectric applications* // Jaime Andrés Pérez Taborda
14:30 - 16:00	Sesiones orales		**TOPICS**
		Aula 19-522	**Metrología**
		Aula 27-101	**Instrumentación científica e industrial**
		Aula 38-222	**Instrumentación científica e industrial**
		Aula 29-203	**Óptica y fotónica**
		Aula 38-118	**Óptica y fotónica**
16:00 - 16:30	Refrigerio	Hall del Auditorio Fundadores // Plazoleta del Auditorio Fundadores	
16:30 - 18:00	Sesiones murales	Hall del Auditorio Fundadores // Plazoleta del Auditorio Fundadores	Presentación de pósters

Jean Paul Allain
University of Illinois at Urbana - USA

Joseph A. Giaime
Observatory Head, LIGO Livingston (Caltech) - Louisiana State University - USA

Miércoles 28 de septiembre
Universidad Nacional - sede Medellín

HORA	ACTIVIDAD	LUGAR	DESCRIPCIÓN
8:30-10:00	Sesiones orales	Aula Máxima (11-202)	**TOPICS** Biofísica
		Aula TIC BL 16 109-3	Biofísica y biotecnología
		Aula 41-102	Nanotecnología y ciencia de los materiales
		Aulas 21-419/ 21-420	Física del estado sólido y aplicaciones
		Aula 21-407	Ciencia de los materiales
10:00 - 10:30	Refrigerio	Cafetería El Carambolo (BL 44 - Ágora)	
10:30 - 11:00	Conversatorio de profesores invitados con estudiantes	Aula Máxima (11-202)	Conversatorio dirigido por el profesor Alcides Montoya
11:00 - 12:00	Sesiones orales	Aula Máxima (11-202)	**TOPICS** Nanotecnología y ciencia de los materiales
		Auditorio Samuel Melguizo (24-307)	Ciencia de los materiales
		Aula 21-407	Geofísica y agricultura de precisión
		Aula 41-102	Física del estado sólido y aplicaciones
		Aulas 21-419/ 21-420	Materia condensada, física del plasma, espectroscopía
12:00 - 14:00	Almuerzo libre		
14:00 - 14:30	Conferencia paralela	Auditorio Samuel Melguizo (24-307)	*Spectrum of classes of point emitters in non-paraxial optical field modeling* // Román Castañeda Sepúlveda
		Aula Máxima (11-202)	*Potencial eléctrico de la membrana externa mitocondrial en el control del metabolismo energético de las células vivas* // Víktor Lemeshko
		Aula 41-102	*Actividades de investigación del grupo de física radiológica* // Jorge Anselmo Puerto Ortiz
14:30 - 16:00	Sesiones orales	Aula Máxima (11-202)	**TOPICS** Nanotecnología y ciencia de los materiales
		Auditorio Samuel Melguizo (24-307)	Física general y enseñanza de la física
		21-407	Ciencia de los materiales
		41-102	Materia condensada
		Aulas 21-419/ 21-420	Física del estado sólido y aplicaciones
16:00 - 16:30	Refrigerio	Cafetería El Carambolo (BL 44 - Ágora)	
16:30 - 18:00	Sesiones orales	Aula Máxima (11-202)	**TOPICS** Nanotecnología y ciencia de los materiales
		Auditorio Samuel Melguizo (24-307)	Emprendimiento científico y tecnológico
		Aula 21-407	Física general y enseñanza de la física
		Aula 41-102	Ciencia de los materiales
		Aulas 21-419/21-420	Materia condensada
18:00 - 20:00	Asamblea SCIF	Aula Máxima (11-202)	

Jueves 29 de septiembre
Universidad EAFIT

HORA	ACTIVIDAD	LUGAR	DESCRIPCIÓN
8:00 - 9:00	Sesiones orales	38-118	**TOPICS** Ciencia de los materiales
		38-119	Energías renovables y no convencionales, instrumentación científica e industrial
		38-121	Física del plasma y aplicaciones
		38-123	Óptica y fotónica
		38-216	Física computacional y laboratorios virtuales
9:00 - 10:00	Conferencia plenaria	Auditorio Fundadores	*Applying Physics to Industrial Challenges in the 21st Century* // Alessandro Curioni
10:00 - 10:30	Refrigerio	Hall del Auditorio Fundadores // Plazoleta del auditorio Fundadores	
10:30 - 11:30	Sesiones orales	27-203	**TOPICS** Ciencia de los materiales
		27-204	Emprendimiento científico y tecnológico
		27-304	Óptica y fotónica
		29-204	Biotecnología
		29-206	Física teórica
11:30 - 14:00	Almuerzo libre		
14:00 - 14:30	Conferencias paralelas	Auditorio Fundadores	*Estudio Monte Carlo del efecto de la interacción dipolar en nanoestructuros magnéticas tubulares* // Johans Restrepo
		38-122	*Uso de la transformada vórtice en métodos de reconstrucción de fase orientado hacia la integración de instrumentos ópticos de la industria aeroespacial* // René Restrepo
		38-222	*Proyecto Helium, historia de un emprendimiento en energía solar* // José Ignacio Marulanda Bernal
14:30 - 16:00	Sesiones orales	19-522	**TOPICS** Ciencia de los materiales
		38-122	Procesamiento de imágenes
		38-222	Ciencia de los materiales
		29-203	Instrumentación científica e industrial, microingeniería
		29-204	Física atmosférica, instrumentación científica e industrial
16:00 - 16:30	Refrigerio	Hall del Auditorio Fundadores // Plazoleta del Auditorio Fundadores	
16:30 - 18:00	Sesiones murales	Hall del Auditorio Fundadores // Plazoleta del Auditorio Fundadores	Presentación de pósters

Viernes 30 de septiembre
Universidad EAFIT

HORA	ACTIVIDAD	LUGAR	DESCRIPCIÓN
8:00 - 9:00	Sesiones orales	19-522	**TOPICS** Ciencia de los materiales
		27-101	Física computacional y laboratorios virtuales
		29-204	Física teórica
		29-205	Metrología y enseñanza de la física
		29-206	Instrumentación científica e industrial
9:00 - 10:00	Conferencia plenaria	Auditorio Fundadores	*Complex Quantum states on a Chip* // Roberto Morandotti
10:00 - 10:30	Refrigerio	Hall del Auditorio Fundadores // Plazoleta del Auditorio Fundadores	
10:30 - 11:30	Conferencia plenaria	Auditorio Fundadores	*Nanoscience and Technology of Layered Materials: From Doping and Controlled Synthesis to sensors and Medical Applications* // Mauricio Terrones
11:30 - 12:00	Clausura	Auditorio Fundadores	Ceremonia de clausura
12:00 - 14:00	Almuerzo libre		
14:00 - 18:00	Tarde cultural libre		

Alessandro Curioni
IBM Fellow, Vice President Europe and Director IBM Research - Zurich Lab.

Mauricio Terrones
The Pennsylvania State University - USA

Roberto Morandotti
Institut national de la recherche scientifique, INRS-EMT - Canada

UNIVERSIDAD EAFIT — SCIF — UNIVERSIDAD NACIONAL DE COLOMBIA

Lunes 26 de septiembre
Universidad Nacional - sede Medellín

HORA	ACTIVIDAD	LUGAR	DESCRIPCIÓN
8:00 - 9:00	Inscripciones	Hall de la biblioteca Efe Gómez	Entrega de escarapelas y material del congreso
9:00 - 10:00*	Minicurso 1	Aula Máxima (11-202)	*Simulación Monte Carlo aplicada al estudio de materiales //* Johans Restrepo
	Minicurso 2	Auditorio Samuel Melguizo (24-307)	*Aerospace electronic instrumentation and systems design applied to aircraft avionics //* Mario Andrés Córdoba Gonzalez
	Minicurso 3	Aula 41-102	*Energía solar fotovoltaica, dimensionamiento de un sistema autosostenible //* José Ignacio Marulanda Bernal
	Minicurso 4	Aula 21-402/ 21-403	*Uso de los dispositivos móviles en los laboratorios de física básica //* Diego Aristizábal
	Minicurso 5	Aula 21-228 (sala 1)	*Fundamentos de simulación Monte Carlo para el estudio de propiedades magnéticas de materiales //* Elisabeth Restrepo Parra
	Minicurso 6	Aula TIC BL 16 109-3	*Principios de microscopía holográfica digital con y sin lentes //* Jorge Iván García Sucerquia
10:00 - 10:30	Refrigerio		Cafetería El carambolo (BL 44 - Ágora)
10:30 - 12:00	Minicursos 1-6		Como se indicó en la franja*
12:00 - 14:00	Almuerzo libre		
14:00 - 15:00	Conferencia plenaria	Auditorio Samuel Melguizo (24-307) // Aula Máxima (11-202)***	*Graphene and related materials: from basic science to applications //* Andrea Ferrari
15:00 - 16:00**	Minicurso 7	Aula 21-402/ 21-403	*Uso de los dispositivos móviles en los laboratorios de física básica //* Diego Aristizábal
	Minicurso 8	Aula 12-204	*LIGO: Astrophysics from precision engineering //* Joseph A. Giaime
16:00- 16:30	Refrigerio		Cafetería El carambolo (BL 44 - Ágora)
16:30 - 18:00	Minicursos 7-8		Como se indicó en la franja**
18:00 - 20:00	Coctel	Ágora	Coctel de bienvenida y presentación de grupo artístico en homenaje a Pink Floyd

***Conferencia plenaria retransmitida vía streaming.

Andrea Ferrari
Director of the Cambridge Graphene Centre - Cambridge University (UK)

Martes 27 de septiembre
Universidad EAFIT

HORA	ACTIVIDAD	LUGAR	DESCRIPCIÓN
7:30 - 8:30	Registro	Auditorio Fundadores	Registro de inscritos y matriculados
8:30 - 9:00	Acto de apertura	Auditorio Fundadores	
9:00 - 10:00	Conferencia inaugural	Auditorio Fundadores	*Observation of gravitational waves from black hole coalescence //* Joseph A. Giaime
10:00 - 10:30	Refrigerio	Hall del Auditorio Fundadores // Plazoleta del Auditorio Fundadores	
10:30 - 11:30	Conferencia plenaria	Auditorio Fundadores	*Directed irradiation synthesis: manipulating matter in nanoscale self-organized systems //* Jean Paul Allain
11:30 - 12:00	Conferencia plenaria	Plazoleta del Auditorio Fundadores	Toma de fotografía del Congreso
12:00 - 14:00	Almuerzo libre		
14:00 - 14:30	Conferencia paralela	Auditorio Fundadores	*Espectroscopía óptica de plasmas producida por láser //* Henry Riascos
		Aula 27-101	*Técnicas de Modelamiento y simulación de propiedades físicas de Materiales //* Elisabeth Restrepo
		Aula 19-522	*Ultra-low thermal conductivities in large-area SiGe nanomeshes for thermoelectric applications //* Jaime Andrés Pérez Taborda
14:30 - 16:00	Sesiones orales		**TOPICS**
		Aula 19-522	**Metrología**
		Aula 27-101	**Instrumentación científica e industrial**
		Aula 38-222	**Instrumentación científica e industrial**
		Aula 29-203	**Óptica y fotónica**
		Aula 38-118	**Óptica y fotónica**
16:00 - 16:30	Refrigerio	Hall del Auditorio Fundadores // Plazoleta del Auditorio Fundadores	
16:30 - 18:00	Sesiones murales	Hall del Auditorio Fundadores // Plazoleta del Auditorio Fundadores	Presentación de pósters

Jean Paul Allain
University of Illinois at Urbana - USA

Joseph A. Giaime
Observatory Head, LIGO Livingston (Caltech) - Louisiana State University - USA

IOP Conf. Series: Journal of Physics: Conf. Series **850** (2017) 011003 doi:10.1088/1742-6596/850/1/011003

Miércoles 28 de septiembre
Universidad Nacional - sede Medellín

HORA	ACTIVIDAD	LUGAR	DESCRIPCIÓN
8:30-10:00	Sesiones orales		**TOPICS**
		Aula Máxima (11-202)	Biofísica
		Aula TIC BL 16 109-3	Biofísica y biotecnología
		Aula 41-102	Nanotecnología y ciencia de los materiales
		Aulas 21-419/ 21-420	Física del estado sólido y aplicaciones
		Aula 21-407	Ciencia de los materiales
10:00 - 10:30	Refrigerio		Cafetería El Carambolo (BL 44 - Ágora)
10:30 - 11:00	Conversatorio de profesores invitados con estudiantes	Aula Máxima (11-202)	Conversatorio dirigido por el profesor Alcides Montoya
11:00 - 12:00	Sesiones orales		**TOPICS**
		Aula Máxima (11-202)	Nanotecnología y ciencia de los materiales
		Auditorio Samuel Melguizo (24-307)	Ciencia de los materiales
		Aula 21-407	Geofísica y agricultura de precisión
		Aula 41-102	Física del estado sólido y aplicaciones
		Aulas 21-419/ 21-420	Materia condensada, física del plasma, espectroscopía
12:00 - 14:00			Almuerzo libre
14:00 - 14:30	Conferencia paralela	Auditorio Samuel Melguizo (24-307)	*Spectrum of classes of point emitters in non-paraxial optical field modeling* // Román Castañeda Sepúlveda
		Aula Máxima (11-202)	*Potencial eléctrico de la membrana externa mitocondrial en el control del metabolismo energético de las células vivas* // Viktor Lemeshko
		Aula 41-102	*Actividades de investigación del grupo de física radiológica* // Jorge Anselmo Puerta Ortiz
14:30 - 16:00	Sesiones orales		**TOPICS**
		Aula Máxima (11-202)	Nanotecnología y ciencia de los materiales
		Auditorio Samuel Melguizo (24-307)	Física general y enseñanza de la física
		21-407	Ciencia de los materiales
		41-102	Materia condensada
		Aulas 21-419/ 21-420	Física del estado sólido y aplicaciones
16:00 - 16:30	Refrigerio		Cafetería El Carambolo (BL 44 - Ágora)
16:30 - 18:00	Sesiones orales		**TOPICS**
		Aula Máxima (11-202)	Nanotecnología y ciencia de los materiales
		Auditorio Samuel Melguizo (24-307)	Emprendimiento científico y tecnológico
		Aula 21-407	Física general y enseñanza de la física
		Aula 41-102	Ciencia de los materiales
		Aulas 21-419/21-420	Materia condensada
18:00 - 20:00	Asamblea SCIF	Aula Máxima (11-202)	

Jueves 29 de septiembre
Universidad EAFIT

HORA	ACTIVIDAD	LUGAR	DESCRIPCIÓN
8:00 - 9:00	Sesiones orales		**TOPICS**
		38-118	Ciencia de los materiales
		38-119	Energías renovables y no convencionales, instrumentación científica e industrial
		38-121	Física del plasma y aplicaciones
		38-123	Óptica y fotónica
		38-216	Física computacional y laboratorios virtuales
9:00 - 10:00	Conferencia plenaria	Auditorio Fundadores	*Applying Physics to Industrial Challenges in the 21st Century* // Alessandro Curioni
10:00 - 10:30	Refrigerio		Hall del Auditorio Fundadores // Plazoleta del auditorio Fundadores
10:30 - 11:30	Sesiones orales		**TOPICS**
		27-203	Ciencia de los materiales
		27-204	Emprendimiento científico y tecnológico
		27-304	Óptica y fotónica
		29-204	Biotecnología
		29-206	Física teórica
11:30 - 14:00			Almuerzo libre
14:00 - 14:30	Conferencias paralelas	Auditorio Fundadores	*Estudio Monte Carlo del efecto de la interacción dipolar en nanoestructuras magnéticas tubulares* // Johans Restrepo
		38-122	*Uso de la transformada vórtice en métodos de reconstrucción de fase orientado hacia la integración de instrumentos ópticos de la industria aeroespacial* // René Restrepo
		38-222	*Proyecto Helium, historia de un emprendimiento en energía solar* // José Ignacio Marulanda Bernal
14:30 - 16:00	Sesiones orales		**TOPICS**
		19-522	Ciencia de los materiales
		38-122	Procesamiento de imágenes
		38-222	Ciencia de los materiales
		29-203	Instrumentación científica e industrial, microingeniería
		29-204	Física atmosférica, instrumentación científica e industrial
16:00 - 16:30	Refrigerio		Hall del Auditorio Fundadores // Plazoleta del Auditorio Fundadores
16:30 - 18:00	Sesiones murales	Hall del Auditorio Fundadores // Plazoleta del Auditorio Fundadores	Presentación de pósters

Viernes 30 de septiembre
Universidad EAFIT

HORA	ACTIVIDAD	LUGAR	DESCRIPCIÓN
8:00 - 9:00	Sesiones orales		**TOPICS**
		19-522	Ciencia de los materiales
		27-101	Física computacional y laboratorios virtuales
		29-204	Física teórica
		29-205	Metrología y enseñanza de la física
		29-206	Instrumentación científica e industrial
9:00 - 10:00	Conferencia plenaria	Auditorio Fundadores	*Complex Quantum states on a Chip* // Roberto Morandotti
10:00 - 10:30	Refrigerio		Hall del Auditorio Fundadores // Plazoleta del Auditorio Fundadores
10:30 - 11:30	Conferencia plenaria	Auditorio Fundadores	*Nanoscience and Technology of Layered Materials: From Doping and Controlled Synthesis to sensors and Medical Applications* // Mauricio Terrones
11:30 - 12:00	Clausura	Auditorio Fundadores	Ceremonia de clausura
12:00 - 14:00			Almuerzo libre
14:00 - 18:00			Tarde cultural libre

Alessandro Curioni
IBM Fellow, Vice President Europe and Director IBM Research - Zurich Lab.

Mauricio Terrones
The Pennsylvania State University - USA

Roberto Morandotti
Institut national de la recherche scientifique, INRS-EMT - Canada

Peer review statement

All papers published in this volume of *Journal of Physics: Conference Series* have been peer reviewed through processes administered by the proceedings Editors. Reviews were conducted by expert referees to the professional and scientific standards expected of a proceedings journal published by IOP Publishing.

Atmospheric turbulence temperature on the laser wavefront properties

J C Contreras López[1], A Ballesteros Díaz[2], O J Tíjaro Rojas[1] and Y Torres Moreno[2].
[1]*Electrical, Electronics and Telecommunications Engineering School*
[2]*Optics and Signal Processing Group, Physics School, Universidad Industrial de Santander, 680002 Bucaramanga, Colombia*
juancontreras9212@gmail.com, ballealex9@gmail.com, otijaro@gmail.com, ytorres@uis.edu.co

Abstract. Temperature is a physical magnitude that if is higher, the refractive index presents more important random fluctuations, which produce a greater distortion in the wavefront and thus a displacement in its centroid. To observe the effect produced by the turbulent medium strongly influenced by temperature on propagation laser beam, we experimented with two variable and controllable temperature systems designed as optical turbulence generators (OTG): a Turbulator and a Parallelepiped glass container. The experimental setup use three CMOS cameras and four temperature sensors spatially distributed to acquire synchronously information of the laser beam wavefront and turbulence temperature, respectively. The acquired information was analyzed with MATLAB® software tool, that it allows to compute the position, in terms of the evolution time, of the laser beam center of mass and their deviations produced by different turbulent conditions generated inside the two manufactured systems. The results were reflected in the statistical analysis of the centroid shifting.
Keywords. Optical Turbulence Generators, Atmospheric Turbulence, Laser Beam, Centroid Beam.

1. Introduction

The earth atmosphere is a medium whose refractive index is close to unity. This condition allows, with slight modifications to the current techniques of propagation of light in vacuum, perform propagation simulations through the atmosphere. The refractive index takes random values in space and time produced by the atmospheric variations; it causes a distortion in the propagation of light beam and its wavefront. For this reason optical systems must overcome the great challenge of reduce the turbulence effects [1].

Research is currently ahead seeking to increase the resolution in imaging and reducing errors in communication systems when the presence of turbulence at low altitudes becomes troublesome [2]. Under these conditions some difficulties have been avoided, as in imaging systems where to elude the turbulence at ground level because is much stronger than any other height, the astronomical observatories are placed on top of the mountains [1].
In the following sections, we show the turbulence characterization using a laser beam that propagates a horizontal distance at ground level (at laboratory scale) and the establishment of the temperature modification effects on the centroid of beam.

2. Systems description

2.1. *Turbulator*

To protect variations other than the temperature variables, a metal tube whose dimensions are illustrated in Figure 1, has inside a resistor gap of dimensions 20 cm long, 3 cm in inner diameter and 3.5 cm in outer diameter. The drilling scheme previously designed for feeding said resistor, provides the temperature control inside the system called "turbulator".

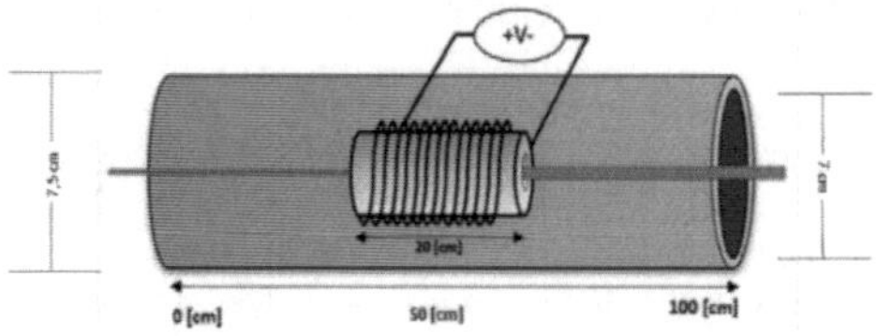

Figure 1. Turbulator, is a temperature controlled environment. Resistor is the hollow cylinder inside.

2.2. *Parallelepiped glass container*

The second design made was a cubic temperature controlled system as shown in Figure 2. Glass room has a hole of 0.8 [cm] diameter on both opposite sides (front and rear) through the laser beam propagates, and is coated to ensure that the light inside the device does not spread to any of the nearest CMOS cameras to reduce its thermal noise.

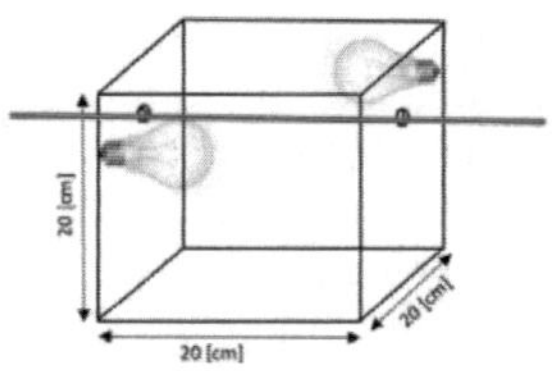

Figure 2. Scheme of parallelepiped glass container.

For wavefront data acquisition, three CMOS cameras and a microcontroller module specialty developed are used (see setup in Figure 3). The cameras CMOS-1 and CMOS-2 are located near the laser mirror output with the idea in mind of estimate the laser intracavity banding effect. The third camera CMOS-3 records such effect in addition to deviations due to turbulence temperature modification along the beam propagation horizontal distance.

Over experimental setup described, acquisition algorithm captures images for each camera and each temperature value on a card expressly developed, in groups of 3000 samples synchronously at 1 Hz of sample frequency. For this reason, acquisition devices (cameras and temperature sensors) are independently connected to a different computer, their operation start from optical synchronization (when laser port is open) with timers preset in each computer where are saved samples to process offline.

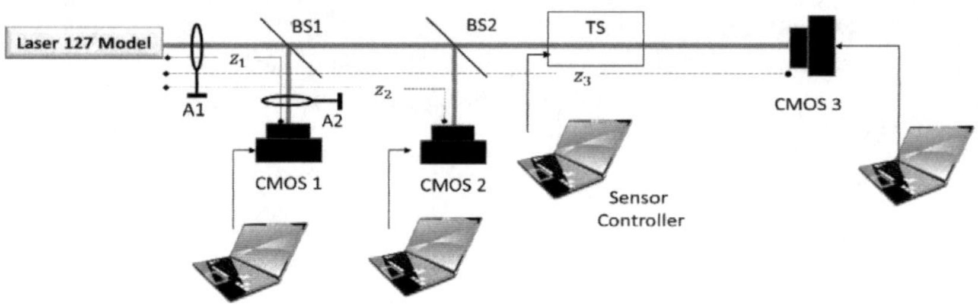

Figure 3. Experimental setup for Turbulator. Z1 = 1.18 m, Z2 = 1.45 m and Z3 = 3.97 m. BS: Beam splitter. CMOS: Sensor. TS: Turbulator implemented (Figure 1 and 2 are replaced according to experiment). A: Attenuator.

2.3. *Temperature distributions*

It was then fundamental to know the temperature distribution within both systems, where it was measured at different spatial positions from the ends to the center (where the highest temperature is

IOP Conf. Series: Journal of Physics: Conf. Series **850** (2017) 012001 doi:10.1088/1742-6596/850/1/012001

concentrated). For this objective, it sets three sensor LM35 and its measurements are shown in Figures 4 to 9.

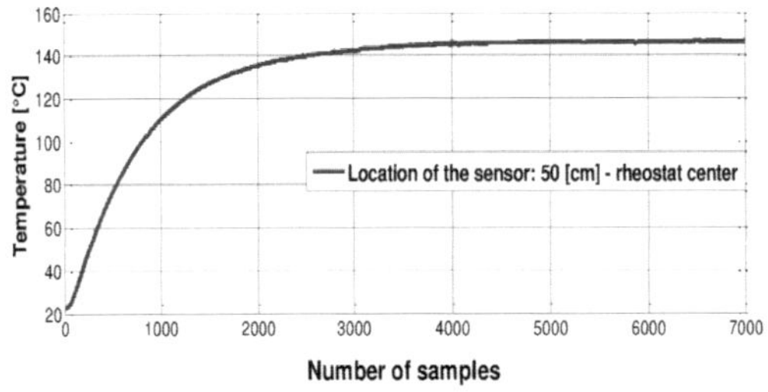

Figure 4. Temperature characteristic curve at the turbulator center. Longitudinal position of sensor, 50 cm at the center of the cross section.

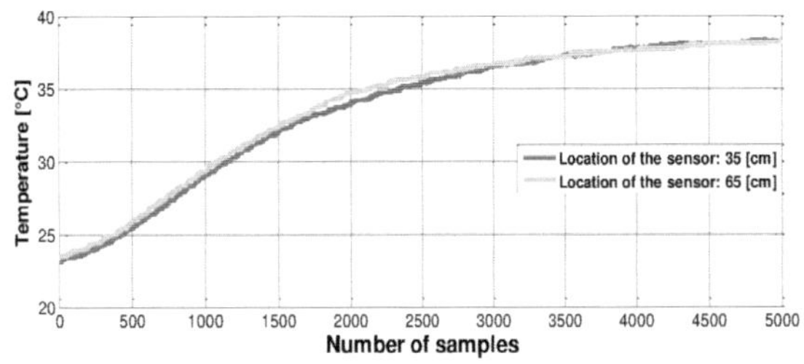

Figure 5. Temperature characteristic curve close to the ends of the turbulator. Longitudinal position of sensors, 35 cm (Orange) and 65 cm (Yellow) at bottom of the cross section.

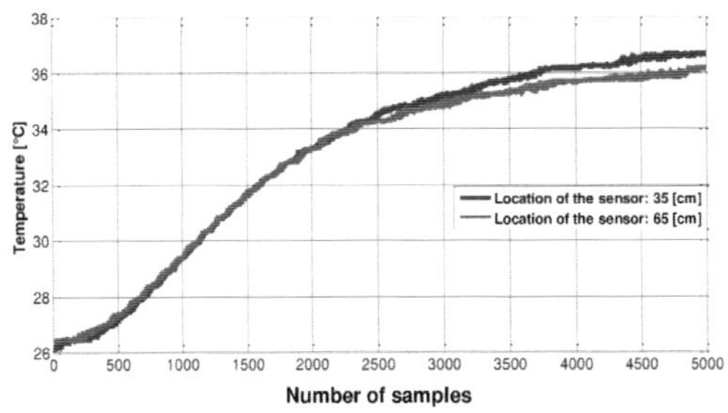

Figure 6. Temperature characteristic curve close to the ends of the turbulator. Longitudinal position of sensors, 35 cm (Blue) and 65 cm (Green) at top of the cross section.

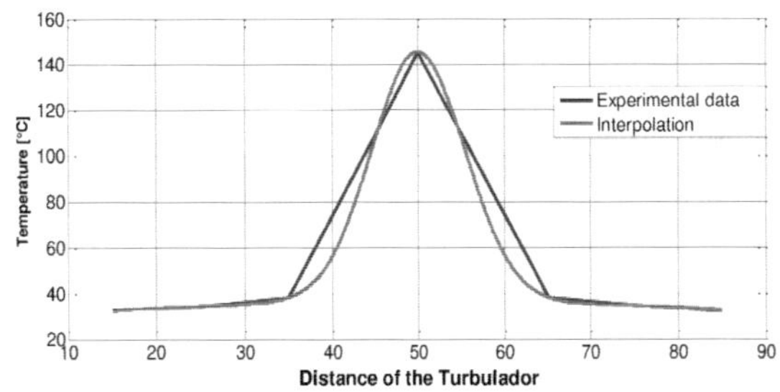

Figure 7. Temperature distribution inside the turbulator according to the distance from the end. Sensors position at bottom and center of the cross section.

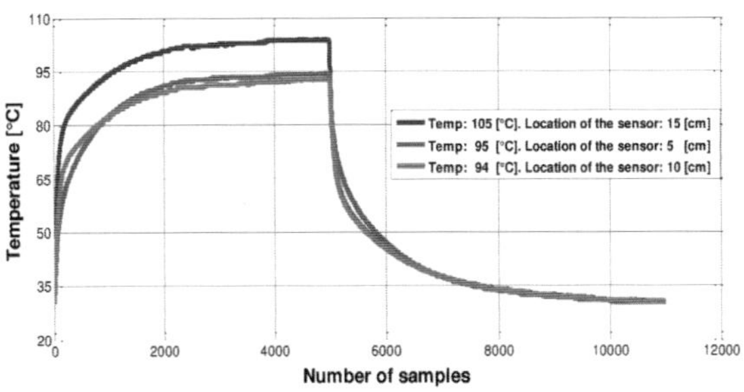

Figure 8. Temperature vs Time characteristic curve in the parallelepiped glass container. Sensors position close to the laser optical path near the ends, as indicated on plot, and at the device center.

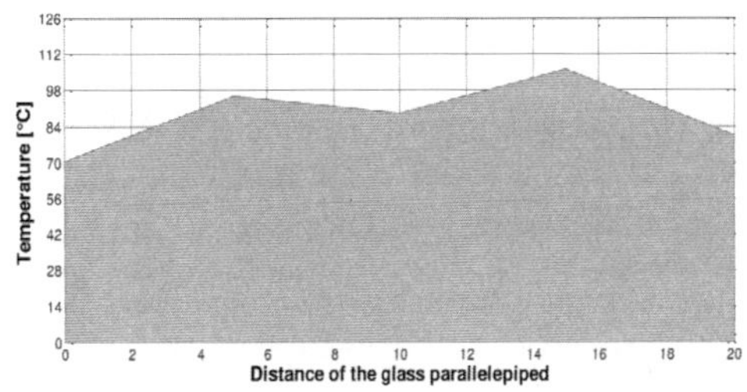

Figure 9. Temperature distribution inside the parallelepiped glass container. Horizontal distance in cm.

Temperature values are selected like to produce significant changes in the captured wavefront at the camera CMOS-3 where the beam centroid becomes fundamental to the turbulence characterization.

3. Results

3.1. Centroid and diameter beam analysis in the laboratory environment.

Figures 10 and 11 correspond to the benchmark made in the natural laboratory environment. Remark that the centroid displacement produced by intracavity beam wander projected from the cameras CMOS-1 and CMOS-2, has similar one to the camera CMOS-3, because the atmospheric turbulence is low under normal laboratory conditions. In fact, CMOS-3 sensor captured directly images whereas that mass center on observation plane was estimated from images of CMOS-1 and CMOS-2.

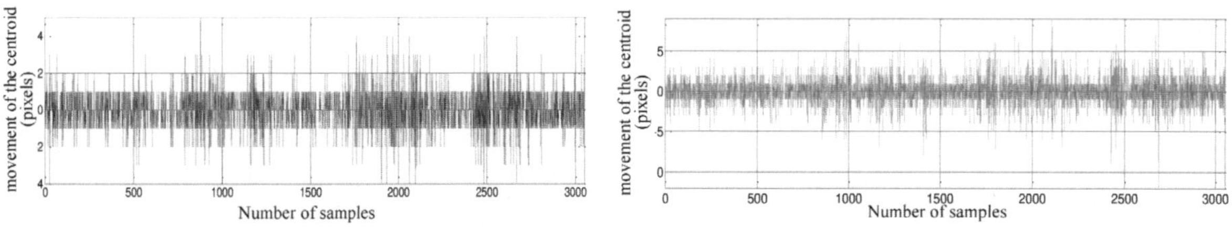

Figure 10. X-shifting of the centroid. Left: Measured on the far camera and, Right: Estimated from nearby cameras.

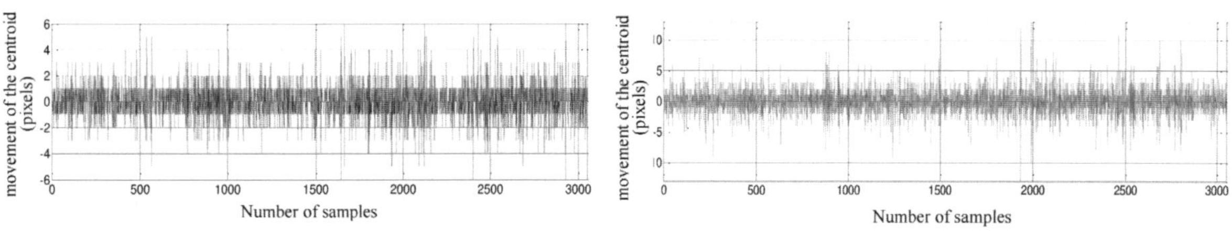

Figure 11. Y-shifting of the centroid. Left: Measured on the far camera and, Right: Estimated from nearby cameras.

The pixel values of the mass center were divided into 34 different positions in the X-axis, and 27 in the Y-axis. The spatial distribution found in each axis was similar, most of the points are found around the fixed coordinate (17,14) with 90 repetitions (close to the center of displacement) as seen in Figure 12.

On the other hand, the beam diameter was also found in order to seek temporal changes confirming that centroids both measured and estimated were consistent (see Figure 13).

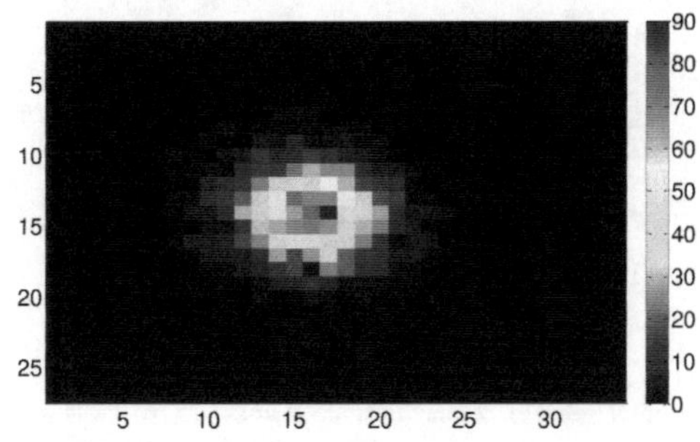

Figure 12. Estimated banding effect on Z3 plane.

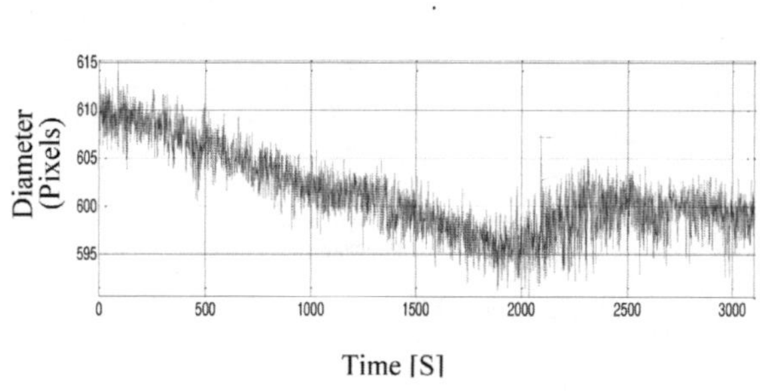

Figure 13. Temporal fluctuations of the beam diameter calculated from experimental data on Camera CMOS-3.

3.2. Beam centroid and its diameter when turbulator is at Temperature of 105 °C.

Establishing a constant internal temperature of 105 °C (schematic is shown in Figure 1), the laser beam propagates through the internal hole diameter of the resistor where the temperature is higher.

Comparable results to shown in Figures 10 to 12, where the data verify the beam behavior and temperature effect were obtained and analyzed. For this experiment, the distance from the laser output mirror to the turbulator was 248 cm and to the camera CMOS-3, 397 cm. The results are shown in Figures 14 to 16, where is evident that the centroid had X and Y shifting like discussed in last section. Initially, the results were unexpected since in the current theoretical models the effect produced by turbulence depend strictly on the temperature [3]. Thus, the results of each test at 105 °C could have been given for the exposed reasons.

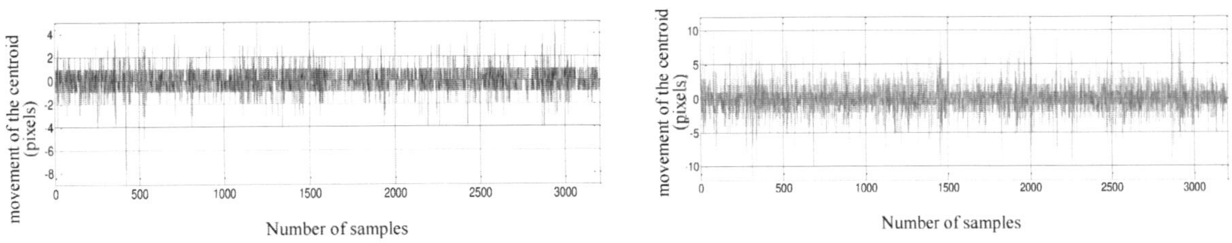

Figure 14. X-shifting of the centroid at 105°C. Left: Measured on the far camera and, Right: Estimated from nearby cameras.

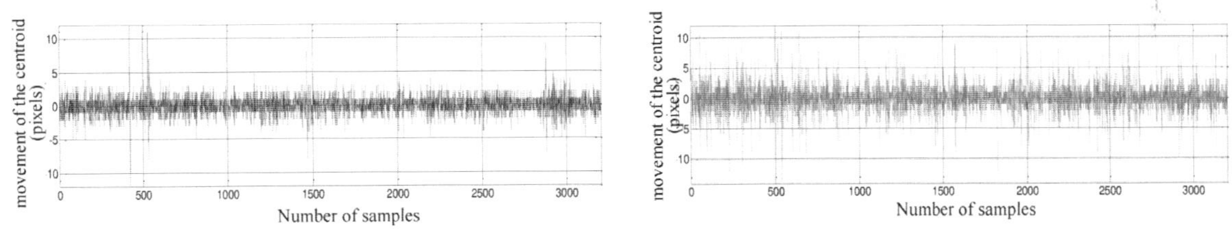

Figure 15. Y-shifting of the centroid at 105°C. Left: Measured on the far camera and, Right: Estimated from nearby cameras.

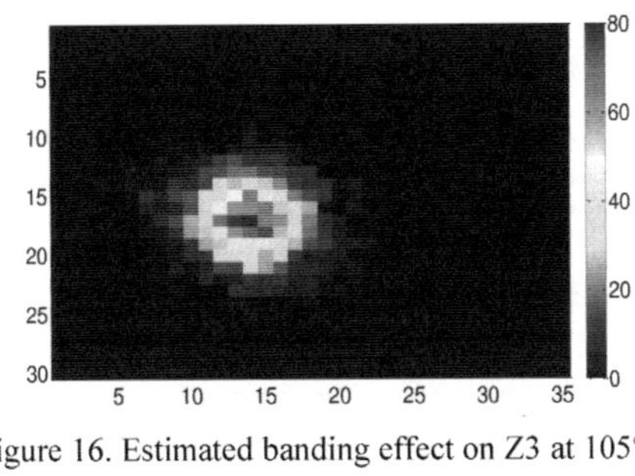

Figure 16. Estimated banding effect on Z3 at 105°C.

3.2.1. The manner in which the temperature source is placed in the medium and generating a uniform temperature gradient.

To explain a little better this phenomenon, an experimental test where the heat source is a point source placed close to the laser beam [4] (measured temperature near to the point source was 210°C) has been conducted. Results presented in figures 17 and 18 have the same results appearance that in sections 3.1 and 3.2, but this time the turbulator has been removed from the setup. In Figure 18, it is provided to observe the difference in the estimated shifting (23 pixels) compared to measured one (10 times larger). This effect is shown in 2D histogram form in Figure 19, wherein the dispersion produced by local temperature variation is evident.

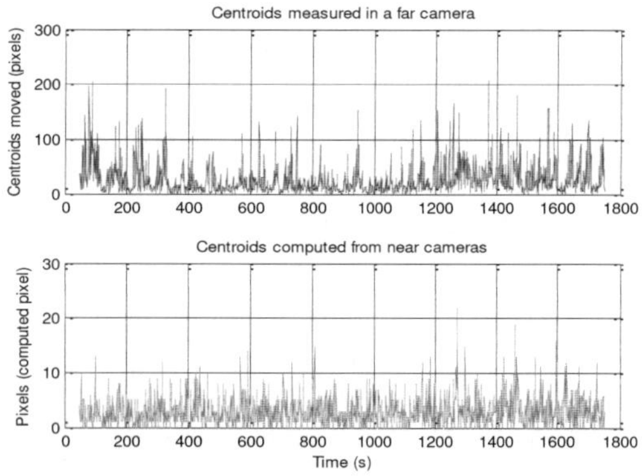

Figure 17. Measured beam centroid on the far sensor CMOS-3 (above) and estimated from nearby cameras (bottom).

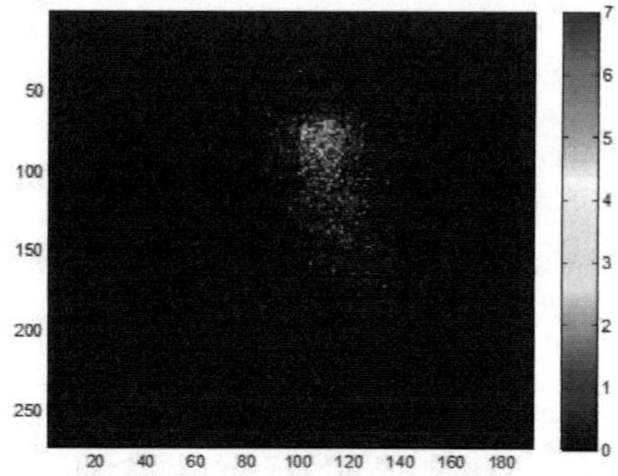

Figure 18. Estimated banding effect on Z3 plane, a point source temperature of 210°C is measured close to the laser beam.

While the resistor remaining surrounded by a larger metal cylinder, the propagation of the laser beam will not be greatly affected by other physical conditions; different temperature distributions can generate greater turbulence like in the case of point heat source. This turbulator configuration ensures a uniform temperature distribution within the device since the temperature gradient is smooth (no any abrupt temperature changes) from the ends of the turbulator and the center thereof, see Figure 19, thus generating less turbulence inside the device.

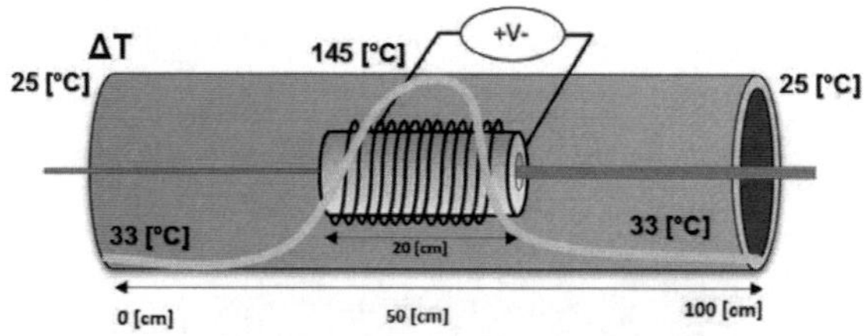

Figure 19. Actual temperature gradient inside the turbulator.

3.2.2. Generation of non-uniform gradients.

To observe the beam centroid behavior against no uniform gradients of temperature, experiments have been conducted in the laboratory using the second device, a parallelepiped glass container (see Figure 20). Due to its geometry and its generation temperature mode is very easy to create no uniform gradients inside and close to its walls.

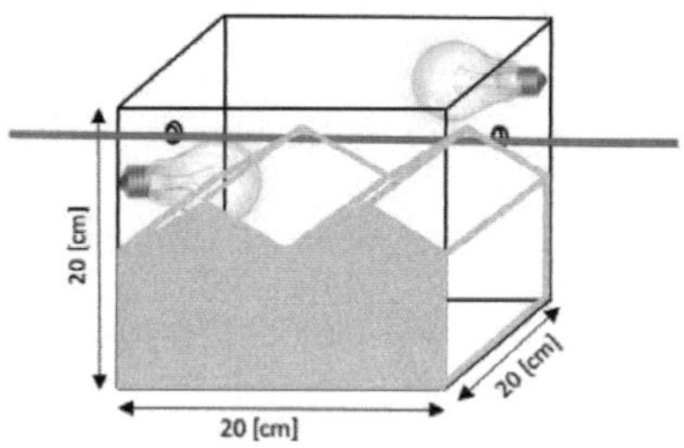

Figure 20. Actual temperature gradient inside the parallelepiped glass container.

IOP Conf. Series: Journal of Physics: Conf. Series **850** (2017) 012001 doi:10.1088/1742-6596/850/1/012001

3.3. Beam centroid and diameter when the parallelepiped glass container is at Temperature of 105 °C.

Tests on the glass parallelepiped shown a notable transversal shifting achieved by the 2D centroid shifting (Figures 21 and 22) the difference in the estimated shifting was 19 pixels compared with the measured one that was 79 pixels. On the X-axis measured shifting was of 54 pixels and the estimated was 13 pixels. On the Y-axis measured shifting was of 79 pixels versus 16 pixels for estimated one, all tests took into account the additive effects of banding and temperature of turbulence.

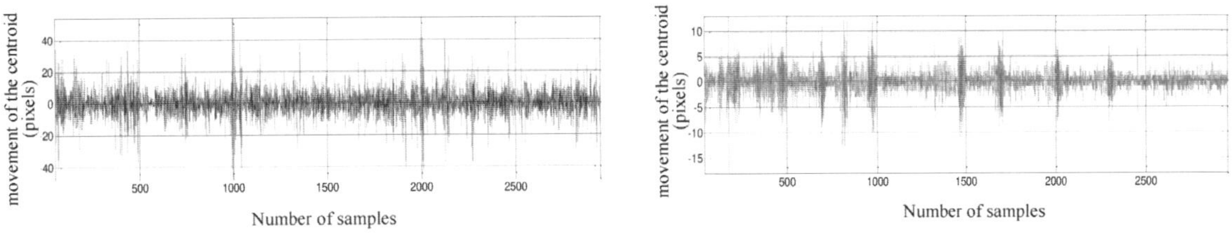

Figure 21. X-shifting of the centroid at 105°C. Left: Measured on the far camera and, Right: Estimated from nearby cameras.

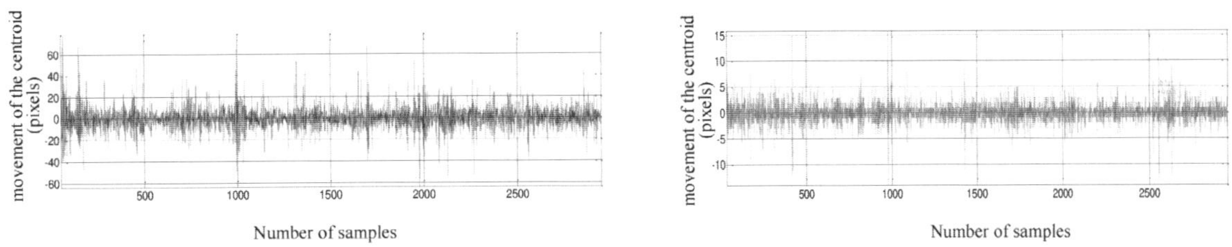

Figure 22. Y-shifting of the centroid at 105°C. Left: Measured on the far camera and, Right: Estimated from nearby cameras.

The pixel values of the mass center were divided into 140 different positions in the X-axis, and 100 for the Y-axis. The distribution was like that found for each axis, most of the points are found around points coordinate (65,51) with a maximum of 15 repetitions at some points of the 2D histogram (near the center of the shifting) as shows Figure 23. In Figure 24 is displayed consistency with the measures of beam diameter.

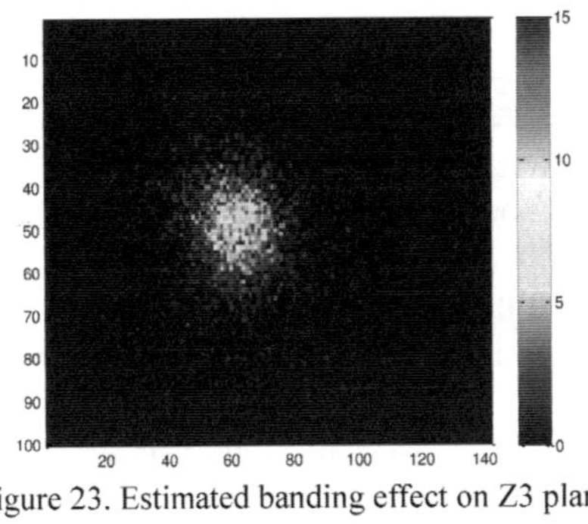

Figure 23. Estimated banding effect on Z3 plane.

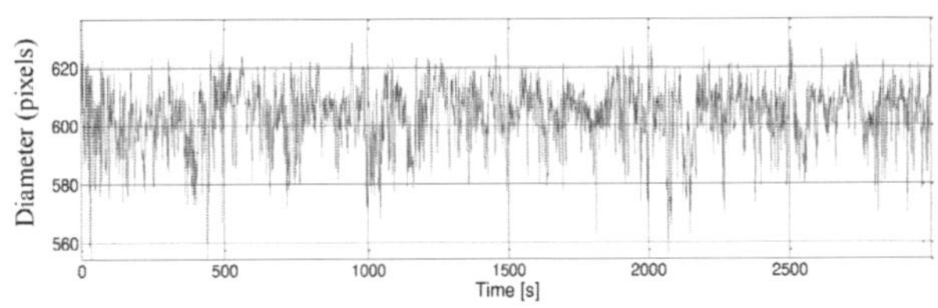

Figure 24. Temporal estimated beam diameter from experimental data on the Camera CMOS-3.

Table 1. Maximum centroid shifting for all experimental tests.

Test (T)		Centroid Movement (pixels) Radially		X axis		Y axis	
		Measured	Calculated	Measured	Calculated	Measured	Calculated
Benchmark		7	13	5	9	6	13
Turbulator	T1	8	10	5	8	7	14
	T2	14	7	6	6	13	7
	T3	14	15	5	12	12	12
	T4	15	15	7	9	14	9
Parallelepiped glass container	T1	17	5	13	5	12	9
	T2	49	10	40	8	43	11
	T3	48	11	33	9	38	17
	T4	79	19	54	13	74	16
Point heat source	T1	23	205	-	-	-	-
Turbulator: propagation distance: 153 cm.		22	25	16	18	16	18

3.4. Turbulence characterization.

3.4.1. Structure function.

In physical optics, correlation functions are widely used to describe statistically varying fields, so we might find it natural to express the fluctuations in such terms; however, since the mean value of the refractive index can vary slowly with position and time, it becomes uncertain which changes should be considered as drifts of the mean and which as fluctuations around their mean value. Kolmogorov devised a slightly different way of expressing his results which overcomes this difficulty. He defined a structure function to relate the values of a function $f(\boldsymbol{r})$ at neighboring points $\boldsymbol{r}_1$ and $\boldsymbol{r}_2$, this is a function only of $\boldsymbol{r} = \boldsymbol{r}_2 - \boldsymbol{r}_1$ and not of their individual values. The atmospheric turbulence is a non-stationary process because its average value varies over time. Therefore, it is necessary to use an expression for interpreting when stationary fluctuations increases. To do this the structure function is used [5].

$$D_f(\boldsymbol{r}) = \langle [f(\boldsymbol{r}_1 + \boldsymbol{r}) - f(\boldsymbol{r}_1)]^2 \rangle \quad (1)$$

where $\langle \dots \rangle$ indicates a time-averaged value in a homogeneous region [6].
Although vertical profiles are commonly known and described in Labeyrie (5.2 section) [6], our model is thought in horizontal path, because empirical models had rebuilt atmosphere layers in vertical profiles. Then, to analyze fluctuations centroid structure function was performed in experimental test performed both in the parallelepiped glass container (see Figures 26 to 28) and in the turbulator on the CMOS-3 camera. Each of these graphs have temporal windows of ten seconds, that means, it has 2800 samples for each test, therefore, 280 values of local fluctuations were obtained.
Since the information obtained since analysis of the structure-function of the tests performed in glass parallelepiped, it was observed that global fluctuations increase to higher temperature levels (Figures 25-26). Global fluctuations centroid on the Y-axis are greater than the X-axis, due to the relative temperature difference which exists at different heights in the parallelepiped.

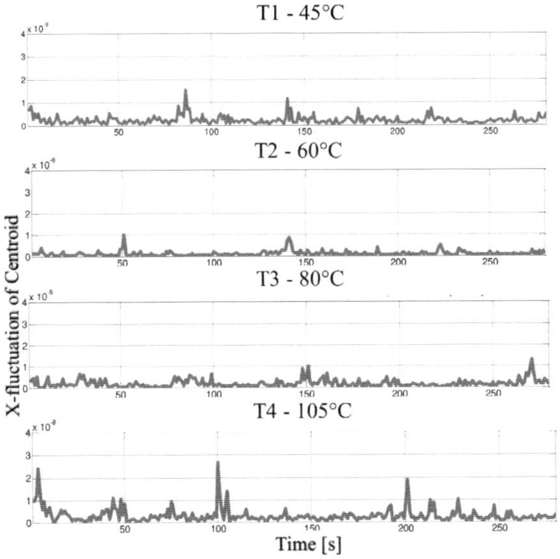

Figure 25. Structure function for component X of centroid, tests T1 to T4 performed in parallelepiped glass container.

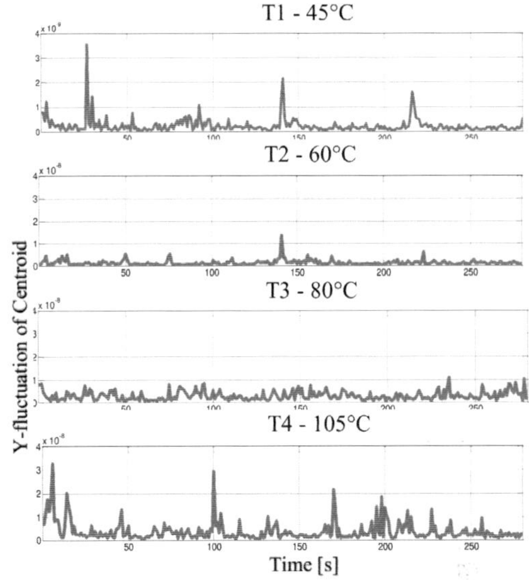

Figure 26. Structure function for component Y of centroid, tests T1 to T4 performed in parallelepiped glass container.

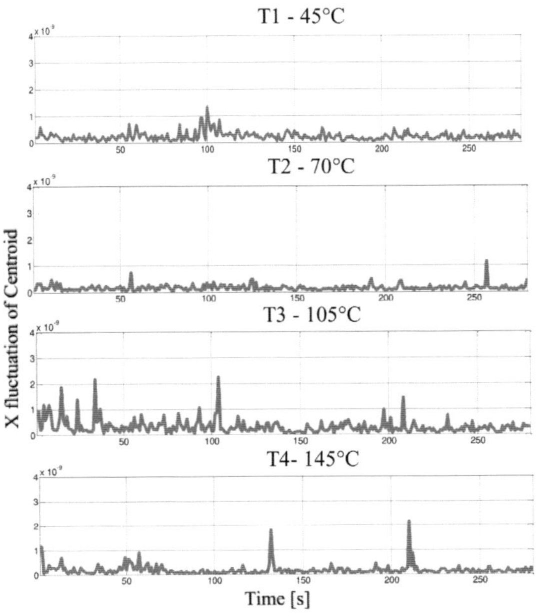

Figure 27. Structure function for component X of centroid, tests T1 to T4 performed in turbulator.

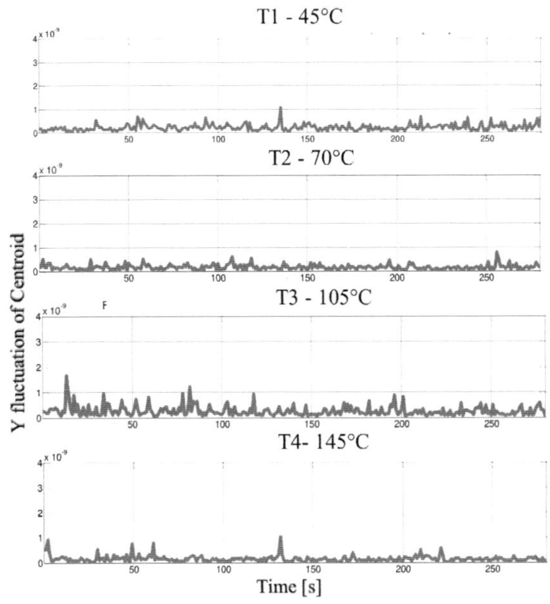

Figure 28. Structure function for component Y of centroid, tests T1 to T4 performed in turbulator.

IOP Conf. Series: Journal of Physics: Conf. Series **850** (2017) 012001 doi:10.1088/1742-6596/850/1/012001

Table 2. Overall average X and Y fluctuation of centroid performed in parallelepiped glass container.

Test	Average of Fluctuation (X-centroid)	Average of Fluctuation (Y-centroid)
T1 – [45 °C]	2.2066e-10	2.4321e-10
T2 – [60 °C]	1.1428e-9	1.3099e-9
T3 – [80 °C]	1.9586e-9	2.8736e-9
T4 – [105 °C]	2.8566e-9	3.6986e-9

From the experimental information obtained since the structure function in the turbulator (see Figures 27 and 28), it remarked that global fluctuations in the first and third test increase, while for the second and fourth ones fall, these results are due probably to the way in which the generating temperature gradient has been previously exposed.

Table 3. Overall average X and Y fluctuation of centroid performed in turbulator.

Tests	Average of Fluctuation (X-Centroid)	Average of Fluctuation (Y-Centroid)
T1 – [45 °C]	2.2079e-10	2.2391e-10
T2 – [70 °C]	1.6591e-10	1.5863e-10
T3 – [105 °C]	2.5580e-10	3.1142e-10
T4 – [145 °C]	1.5611e-10	1.8231e-10

Table 4. Overall average X and Y fluctuation of centroid benchmark performed.

Test	Average of Fluctuation
X-Centroid	1.9011e-10
Y-Centroid	2.7546e-10

3.4.2. Inner scale calculation.

The inner scale of turbulence l_0, being the scale at which energy is dissipated by atmospheric turbulence convection, is of fundamental importance in determining surface fluxes of heat and momentum [7].

$$l_0 = 1.08L\left(\frac{\langle \alpha^2 \rangle}{\sigma_I^2}\right)^{1/2} \tag{2}$$

Where α and σ_I^2 are the arrival angle of beam and scintillation index, respectively (see equation 4 for scintillation index). We check out that α is very small (micro-radians order). In addition, computed inner scale are shown in Figures 29 and 30 and Table 5.

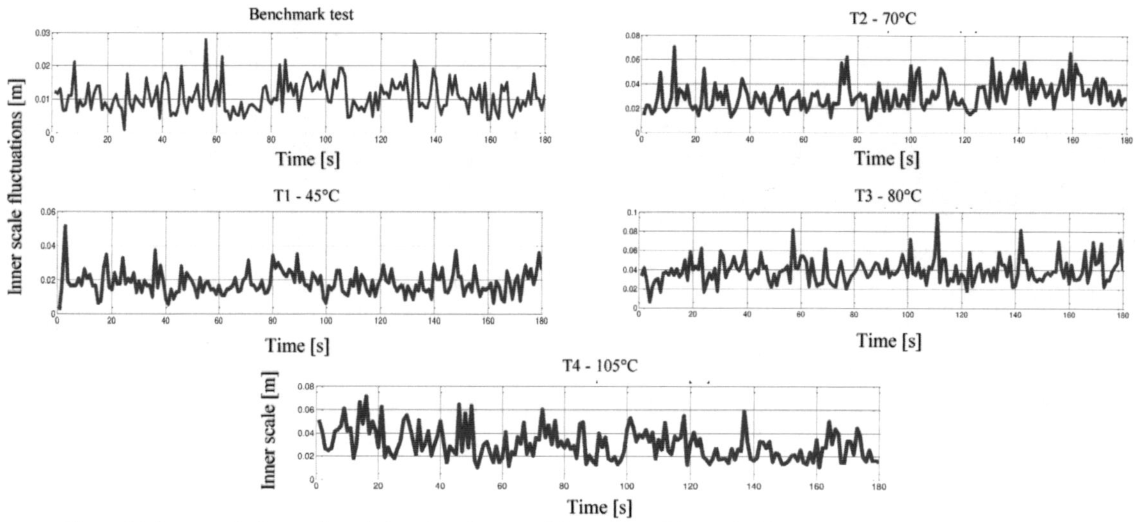

Figure 29. Inner scale fluctuations in benchmark test and from T1 to T4 tests inside the parallelepiped glass container.

Table 5. Inner scale average for benchmark test and from T1 to T4 tests inside the parallelepiped glass container.

Tests	Average of Inner Scale [x10^{-5} m]
Benchmark	1.07
T1 – [45 °C]	1.83
T2 – [60 °C]	3.08
T3 – [80 °C]	3.92
T4 – [105 °C]	3.02

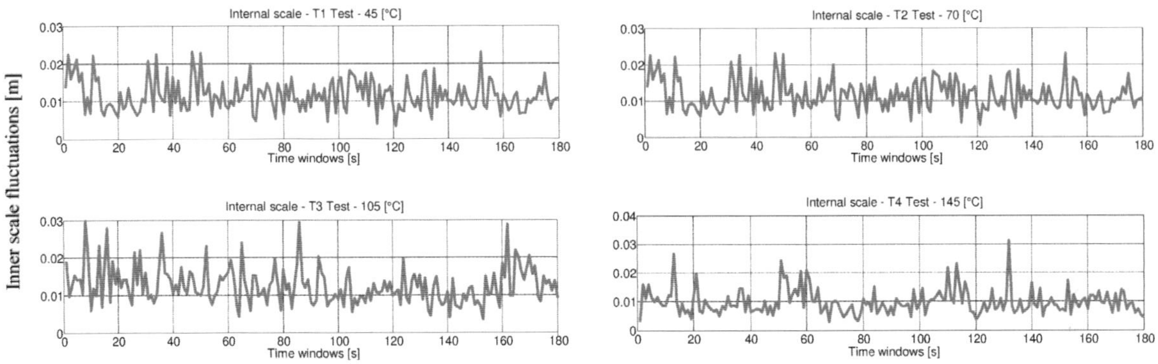

Figure 30. Inner scale fluctuations from T1 to T4 tests inside the turbulador.

Table 6. Inner scale average fluctuation from T1 to T4 tests inside the turbulador.

Tests	Average of Inner Scale [x10^{-5}m]
T1 – [45 °C]	1.10
T2 – [70 °C]	1.16
T3 – [105 °C]	1.26
T4 – [145 °C]	0.98

3.4.3. C_n^2 calculation.

Supposing a plane wave that propagates through the homogeneous and isotropic turbulence, due to we acquire data in a small sensor, within the limits of geometrical optics approximation, that means for $L \ll l_0^2/\lambda$, where L is the path length, l_0 the inner scale and λ the wavelength, the mean square fluctuations of the arrival angle is given by [7]:

$$\langle\alpha^2\rangle = \langle\beta^2\rangle = 3.28 C_n^2 L l_0^{-1/3} \quad (3)$$

where α and β are orthogonal components of the arrival angle, respectively. Their variances can be obtained as derivatives of the phase structure function. Inner scale l_o has been computed in section 3.4.2. Tables 7 - 8, and Figures 31 - 32 show these outcomes.

Table 7. C_n^2 average fluctuation for benchmark test and from T1 to T4 tests inside the parallelepiped glass container.

Tests	Average of C_n^2 [$m^{-\frac{2}{3}}$]
Benchmark	2.0352e-13
T1 – [45 °C]	1.1982e-12
T2 – [60 °C]	8.8053e-12
T3 – [80 °C]	2.0655e-11
T4 – [105 °C]	2.6309e-11

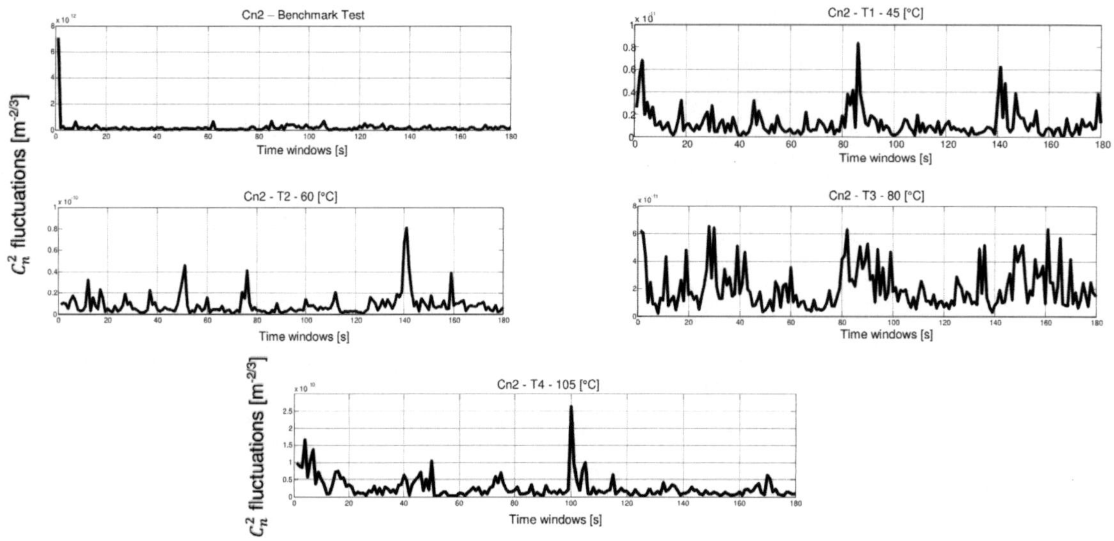

Figure 31. C_n^2 fluctuations from T1 to T4 tests inside the parallelepiped glass container.

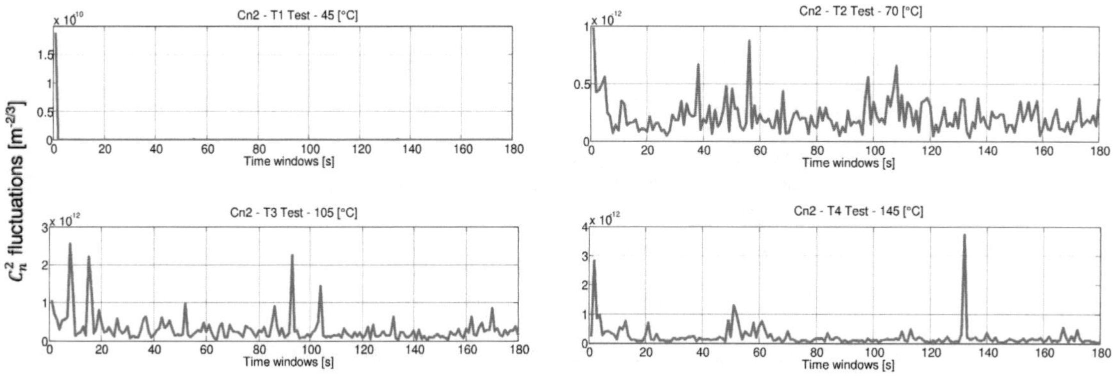

Figure 32. C_n^2 fluctuations from T1 to T4 tests inside the turbulador.

Table 8. C_n^2 average from T1 to T4 tests inside the turbulador.

Tests	Average of C_n^2 $[m^{-\frac{2}{3}}]$
T1 – [45 °C]	1.3196e-12
T2 – [70 °C]	2.1470e-13
T3 – [105 °C]	3.0767e-13
T4 – [145 °C]	2.3043e-13

3.4.4. Scintillation index calculation.

The mean square of the normalized intensity fluctuations is known as scintillation index and is expressed by (7):

$$\sigma_I^2 = \frac{\langle I^2 \rangle - \langle I \rangle^2}{\langle I \rangle^2} \quad (4)$$

This measurement produces an estimation about intensity variance to assess if spot acquired was inside 2D sensor. Finally, we show in Figures 33-34 and Tables 9 -10 scintillation index.

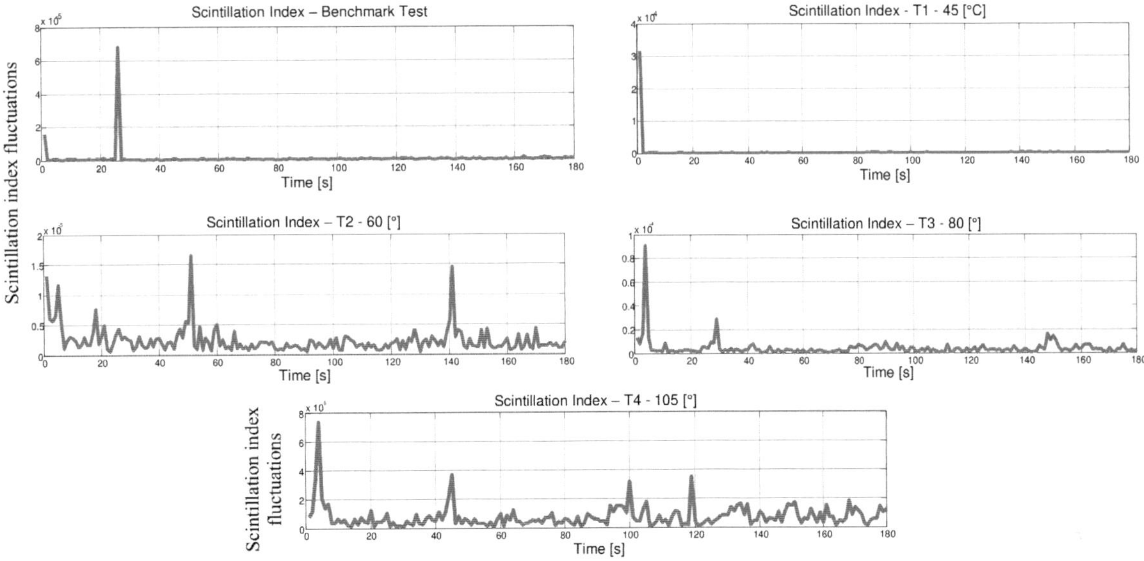

Figure 33. Scintillation index fluctuations in benchmark test and from T1 to T4 inside the parallelepiped glass container.

Table 9. Scintillation index average for benchmark test and from T1 to T4 inside the parallelepiped glass container.

Tests	Average of Scintillation
Benchmark	1.0001e-06
T1 – [45 °C]	2.7890e-06
T2 – [60 °C]	2.3425e-06
T3 – [80 °C]	4.2129e-06
T4 – [105 °C]	7.9181e-06

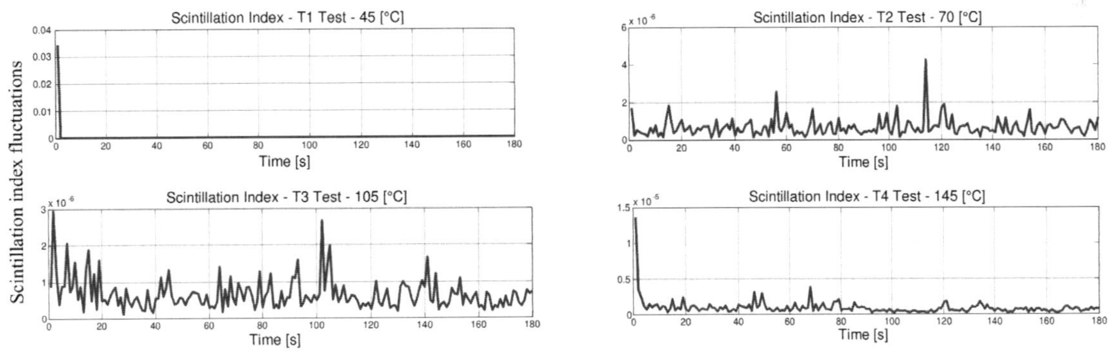

Figure 34. Scintillation index fluctuations from T1 to T4 inside the turbulator.

Table 10. Scintillation index from T1 to T4 inside the turbulator.

Tests	Average of Scintillation
T1 – [45 °C]	1.9161e-04
T2 – [70 °C]	6.6046e-07
T3 – [105 °C]	6.5561e-07
T4 – [145 °C]	9.1303e-07

4. Conclusions

The centroid behavior of the laser beam in the experimental tests performed inside the turbulator had similar results to the test conducted in the atmosphere at normal conditions inside the lab; we suggest as explanation the way in which the transfer of heat from the source to the surrounding air and as the gradient of uniform temperature is produced. The inner scale had a similar performance.

The centroid transversal shifts present greater fluctuations along its two orthogonal coordinate axis, it is due to the positive gradient of temperature that exists inside the turbulator throughout its height.

The structure function shows big fluctuations when centroid transversal shifts increases at higher temperature levels, it is easy to see in the parallelepiped glass container. While in the turbulator it was observed that these fluctuations were presented with strong intensity in the first and third experimental test, while in the second and fourth decreases with respect to the above ones without one good defined behavior pattern.

We show that is possible to calculate the inner scale, C_n^2 and scintillation index of generated turbulence using a very simple method based in the measurement and analysis of the propagated beam wavefront.

A possible deduction from experimental results is concerning the inner scale and C_n^2 at high temperature in turbulator prototype, in our case, fluctuations were similar to ones under the normal conditions in laboratory then, we suggest that dependences between the inner and outer scale did not allow heat convection and, its consequences were small shifts of beam centroid. Future works would yield to identify this hypothesis, we must orient tests with other devices patterns.

5. Acknowledgments

Research partially supported by Universidad Industrial de Santander with project 5707. Thanks to Colciencias for financial support to project 110256933773: “Uso de la telescopía de Fourier de tiempo promedio para caracterizar la turbulencia horizontal a baja altura”, National Call for the Bank of Projects in Science, Technology and Innovation 2012. O.J. Tíjaro Rojas acknowledges the support and sponsorship from Colciencias, by Call 647-2014: National Doctoral program.

6. References

[1] Schmidt, J.D. "*Numerical Simulation of Optical Wave Propagation whit examples in MATLAB*," Bellingham, Washington USA: SPIE PRESS. 2010.

[2] Torres Moreno, Y. "*Uso de la telescopía de Fourier de tiempo promedio para caracterizar la turbulencia horizontal a baja altura*," (Colciencias-UIS project). Colombia: Grupo de Óptica y Tratamiento de Señales, Universidad industrial de Santander. 2012.

[3] Stewart T. "*General Theory of Light Propagation and Imaging through the Atmosphere,*" Volume 196. Scotland, UK: Springer Series in Optical Sciences, pp 48-50, 2016

[4] Tíjaro Rojas O.J., Galeano Traslaviña Y.A., Torres Moreno Y. *"Método para medir los efectos de la turbulencia,"* in ENO-CANCOA 2015, Cali, Colombia, 2015.

[5] Funes G., Trabajo de grado: "*Determinación de la constante de estructura del aire turbulento mediante interferometría Young,"* Buenos Aires, Argentina: Universidad Nacional de la Plata, 2011.

[6] Labeyrie, A., Lipson, S. G., & Nisenson, P. "*An introduction to optical stellar interferometry*," Cambridge University Press. 2006.

[7] Consortini, A., Sun, Y. Y., Innocenti, C., & Li, Z. P. "Measuring inner scale of atmospheric turbulence by angle of arrival and scintillation," *Optics Communications*, *216* (1), 19-23. 2003.

IOP Conf. Series: Journal of Physics: Conf. Series **850** (2017) 012002 doi:10.1088/1742-6596/850/1/012002

AuNx stabilization with interstitial nitrogen atoms: A Density Functional Theory Study

J. H. Quintero[1], R. Gonzalez-Hernandez[2], R. Ospina[3], A. Mariño[4]
[1] Materiales Nanoestructurados y Biomodelación. Universidad de Medellín, Medellín, Colombia.
[2]Grupo de Investigación en Física Aplicada. Universidad del Norte, Barranquilla, Colombia
[3] Escuela de Física. Centro de Materiales y Nanociencia. Universidad Industrial de Santander, Bucaramanga, Colombia.
[4] Laboratorio de Superconductividad y Nuevos Materiales, Universidad Nacional de Colombia, Bogotá D.C., Colombia

E-mail: jhquintero@udem.edu.co

Abstract. Researchers have been studying 4d and 5d Series Transition Metal Nitrides lately as a result of the experimental production of AuN, PtN, CuN. In this paper, we used the Density Functional Theory (DFT) implementing a pseudopotential plane-wave method to study the incorporation of nitrogen atoms in the face-centered cube (fcc) lattice of gold (Au). First, we took the fcc structure of gold, and gradually located the nitrogen atoms in tetrahedral (TH) and octahedral (OH) interstitial sites. AuN stabilized in: 2OH (30%), 4OH and 4TH (50%), 4OH – 2TH (close to the wurtzite structure) and 6TH (60%). This leads us to think that AuN behaves like a Transition Metal Nitride since the nitrogen atoms look for tetrahedral sites.

Key Words: Crystal Structure, Point Defects, Solid Solutions, Computer Simulation, Superlattices, Nitrides.

1. INTRODUCTION

Transition Metal Nitrides (TMN) are hard materials which are resistant to wear, have high melting points and good thermic stability. They have a significant number of applications in cutting tools, resistant surfaces, semiconductors, or conductivity among others [1-5]. Long time ago (approx. 50 years), the most studied nitrides were 3d series interstitial nitrides [6-9], but after the experimental production of some TMN as Ag_xN, Au_xN, Pt_xN, C_xN in these last ten years, they begun to study strongly 4d and 5d Series Transition Metal Nitrides (Ru, Rh, Pd, Ag, Os, Ir, Pt, and Au) [10-14]. It is interesting to know that these noble metals are part of new TMN family (IrN_2, AgN_2, AuN_2, PtN_2) in which nitrogen atoms settle themselves into interstitial sites of the fcc structure of a metal [15]. Authors as Yu, Kanoun, Zhao, and Mohammed have conducted research on these types of nitrides [16-19]. Zhao considered six structures (Rocksalt, Zinc Blenda, CsCl, WurtZite, NiAs, and WC) using first principle methods to study the electronic and structural properties of nitrides from LaN to AuN. The most stable structures found were Rocksalt for LaN, HfN and AuN, WC for TaN, NiAs and Wurtzite for ReN, OsN, IrN and PtN. They are metallic structures and have covalent bonds. The pressure produced between 5 to 40 GPa demonstrated that as coatings these nitrides may potentially become super hard materials. Kanoun confirm conductive nature and found Rock Salt structure to be the most stable for AuN using DFT [17]. Yu observed the structural stability of this family in fluorite-type structures (IrN_2, PtN_2 y AuN_2) and reported that nitrogen atoms fill in the interstitial sites of a metal structure [15]. Mohammed identified that the most stable form of gold nitride is Au_3N using DFT, his report is based in the comparison of 20 structures between Au_3N, Au_2N and AuN, and observed the phase Au_2N has lower symmetry than phase Au_3N (D09) [20]. Due to the difficulty these nitrides have had being synthetized, there is little research work on their experimental production. The problem today is that we do not know in what phase they are in, so this has led to a lack of databases for most of them for X-Ray Diffraction (XRD); just in the

XRD we found, one can not find notable differences with the ones found for a lattice of pure metal. This is mainly because the structures have the same metal sub-lattice, and in heavy-atom metal dispersion, they are predominant in X- Ray Diffraction (XRD) pattern intensity [16]. By means of ab–initio pseudopotential calculations, Siller has shown a triclinic structure for an Au_3N phase [11]. For XRD, it is observed an approximate 0.1 % dilatation for the fcc structure of Au incorporating Nitrogen atoms, and obtained AuN using techniques as ion sputtering and plasma etching as solid solutions [10, 21]. Quintero reported a species of $AuN_{0.6}$ in an N1s at 398.1 eV narrow spectrum as the first AuN film produced using a pulsed arc technique [13, 22]. In this paper, we present the incorporation of nitrogen atoms to the fcc structure of gold by means of DFT, at interstitial sites (OH and TH). Nitrogen atoms were gradually introduced in ranks of 30, 50 and 60 %, to observe if gold nitride has a TMN behavior, for different M_xN_y (M: Metal, N: Nitrogen) compositions, that lead to mononitrides production with different structures. The most common and important MN, has a basic fcc structure. M_2N nitrides follow it in importance, and they are usually found in hexagonal close packing (hcp), and rhombohedral and orthorhombic structures are formed where there is a larger amount of nitrogen than of a metal [23]. Finally, we determined the possible interstitial sites that nitrogen atoms may take in an fcc gold lattice.

2. COMPUTATIONAL DETAILS

Nitrogen positions in a cubic Au lattice were determined theoretically based on DFT. Energy calculations were conducted using the pseudopotential plane-wave method. The exchange and correlation effects were treated with a Generalized Gradient Approximation (GGA) implemented in Perdew-Wang approximations (PW91) [24]. We used ultrasoft pseudopotentials where states 5d for gold were introduced as valence electrons [25]. Calculations were developed using a QUANTUM-espresso simulation packet [26]. Electron wave functions were expanded up to a cutting energy of 35, and 420 Ry for electronic density. We used a 10×10×10 k-point grid centered on Γ for an irreducible Brillouin zone sampling in the special Monkhorst–Pack schema [27]. We adopted Methfessel-Paxton technique with a 0.02 Ry width value for smearing [28]. Those parameters assure a convergence for total energy close to 5 meV. Since fcc structures have eight octahedral sites (OH) and four tetragonal (TH) sites (Fig 1), different structure configurations were taken for different nitrogen concentrations within the fcc gold lattice. When four nitrogen atoms fill the OH sites (Fig. 1b), which are 6-coordinate, the structure presents a point symmetry of 4/m 3* 2/m. In this notation, '4' stands for the presence of 4-fold rotation axes, 'm' stands for mirror planes, '2' stands for 2-fold rotation axes, and '3*' stands for 3-fold rotoinversion axes. When eight nitrogen atoms are located in the TH sites, which are 4-coordinate, the structure takes also a symmetry 4/m 3* 2/m. In this work, we have located different number of N atoms in the possible OH and TH sites and DFT relaxations were performed by fully relaxing the coordinates of the atoms until the forces were smaller than 0.1 eV/Å. In addition, we resolved Kohn-Sham equations via a self-consistent approach for different volumes generating a characteristic Energy vs Volume curve for each configuration, and we were able to observe how this new TMN stabilizes for different configurations and concentrations.

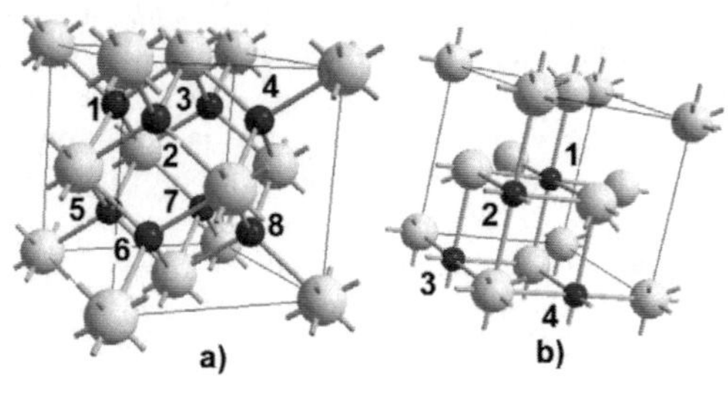

Figure 1. Positions of the interstitial sites, a) Tetrahedral Sites and b) Octahedral Sites.

3. RESULTS AND DISCUSION

For a structural study, we adjusted the energy curves against volume (see figure 2) using Murnaghan's equation of state [29]; and this way, we found a lattice parameter (a), an equilibrium volume (V_0) and a volume module (B). The values found in the fcc gold structure were: a=4.15 Å, V_0=120.6 Å and B=1.73 Mbar, which were very close to the experimental parameters, a=4.08 Å and B=1.70 Mbar. Thus, our result slightly overestimates the experimental values by ~1.7%. In figure 1, one can see the fcc gold structure used as a base to introduce nitrogen atoms in interstitial sites, and later it was easy to determine the structural parameters of each proposed phase. As a result of the different concentrations of nitrogen atoms and of the interstitial positions of a cubic gold lattice, the possible configurations studied are listed on table 1, and figure 3 are shown the most stable structures for 30, 50 and 60 %. Then, for each structure, we found total energy, total force and stress for each of the previously mentioned configurations in order to determine the most stable positions (minimum energy and zero total-force) of the interstitial sites nitrogen atoms take (Table 1).

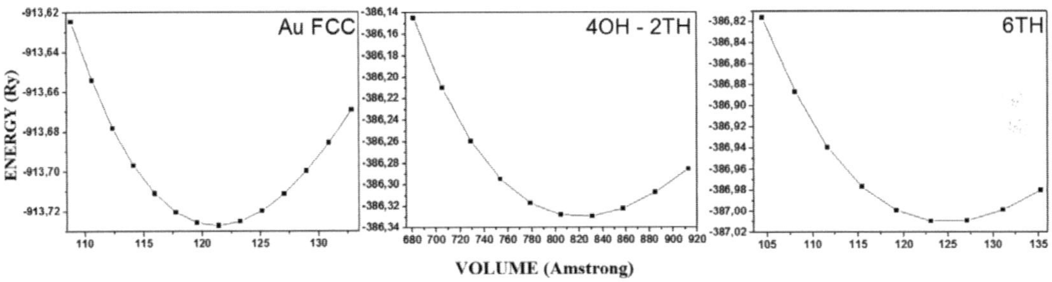

Figure 2. Energy vs Volume Curve for Au-fcc, 4OH–2TH and 6TH structures.

In table 1, nitrogen atoms prefer OH interstitial sites for low nitrogen contents because for 30% of the most stable configuration is 2OH (minimum energy and zero total-force, see figure 3a). For configurations belonging to 50%, the most stable 4OH and 4TH do not present any considerable difference; nevertheless, configuration 4OH presents a slightly lower energy (figure 3b). Zhao et al reported a similar behavior for others TMN [18]. In the literature, one can find that the most stable structure for AuN is rock salt (OH sites) followed in order by zinc blenda, CsCl and wurtzite structures, which have a small energy differences among them [19]. Therefore, it is observed that when nitrogen concentration increases, after the octahedral site are occupied, nitrogen atoms locate into tetrahedral sites trying to imitate the tetrahedral symmetry as it is found in the wurtzite structure [19].

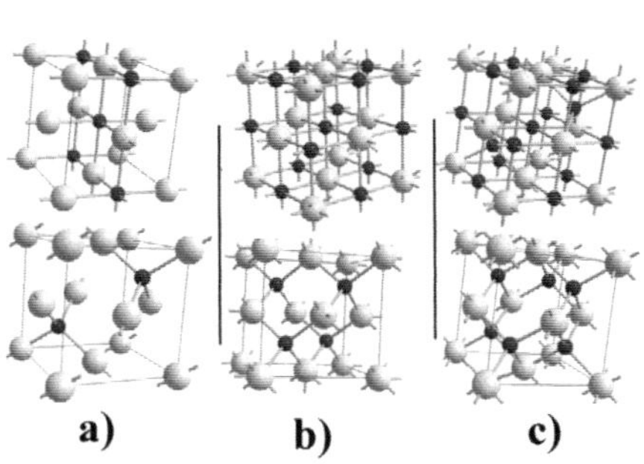

Figure 3. Configurations more stable. a) 30%, b) 50% and c) 60%

For a 60% concentration, the most stable result corresponds to configurations 4OH–2TH and 6TH; in the ones with zero total-force, configuration 4OH–2TH presents minimum energy (figure 3c). For configurations 4OH–2TH and 6TH, we performed lattice relaxation by means of E vs V (figure 1), and determined that the most stable configuration is 6TH. These data agree with what was reported for the TMNs, for they present cubic structures in a range of 30 at 70% [22, 30]; besides, this behavior shows that AuN behaves as a TMN, for when it exceeds this concentration, it may have a phase transition to hexagonal close packing (HCP).

Table 1. Configurations for nitrogen atoms in the Au-fcc structure.

% Nitrogen Atoms		Configuration	OH site	TH site	Energy (Ry)	Total Force (Ry/a.u.)	Stress (kbar)
30%	a	1 OH – 1 TH	5	1	-307.210	0.784	996.29
	b	**2 OH (top fig 3a)**	1, 4	-	**-307.579**	**0**	**646.91**
	c	2 TH	-	5, 7	-307.070	1.131	1179.01
		2 TH	-	5, 3	-307.154	0.825	1174.78
		2 TH (down fig 3a)	-	5, 4	**-307.410**	**0**	**1121.13**
50%	a	**4 OH (top fig 3b)**	1, 2, 3, 4	-	**-347.331**	**0**	**-197.16**
	b	**4 TH (down fig 3b)**	-	6, 1, 7, 4	**-347.312**	**0**	**295.41**
60%	a	1 OH – 5 TH	1	1, 4, 5, 6, 7	-383.296	0,745	3863.15
	b	2 OH – 4 TH	1, 4	6, 1, 7, 4	-383.115	0,453	3803.98
		2 OH – 4 TH	2, 4	6, 1, 7, 4	-383.115	0,453	3804.00
	c	3 OH – 3 TH	2, 3, 4	3, 4, 5	-383.706	0,901	3421.89
	d	4 OH – 2 TH	1, 2, 3, 4	5, 7	-383.783	1,376	3246.57
		4 OH – 2 TH	1, 2, 3, 4	5, 3	-383.842	1,058	3213.93
		4 OH – 2 TH (top fig 3b)	1, 2, 3, 4	5, 4	**-384.213**	**0**	**3029.81**
	e	6 TH	-	1, 3, 4, 5, 6, 7	-383.662	0,852	3862.94
		6 TH (down fig 3c)	-	1, 3, 4, 5, 6, 8	**-384.148**	**0**	**3637.94**

Yu et al reported that new TMN families, among which we can find AuN, are structures that prefer tetrahedral sites for concentrations over 50% [15]. In the literature, Siller and Quintero show that nitrogen is found interstitially in the fcc structure of gold, althoug XRD analysis do not show any parameter changes in a gold lattice [21, 31, 32]. In that case, DFT results without relaxation present a 4.15Å lattice parameter which is close to the one Devia observed experimentally (4.08 Å fcc gold lattice). Also, after structural relaxation, the lattice parameters increase up to 4.956 Å (for configuration 4OH-2TH) and 4.664 Å (for configuration 6TH). Since lattice parameters are quite similar to the results obtained experimentally [31,32], we could think that the structure obtained experimentally could correspond to configuration 4OH–2TH without relaxation, and this is very close to a rock salt structure (figure. 3c).

4. CONCLUSIONS

We used Density Functional Theory (DFT) calculations in order to analyze nitrogen atom interstitial incorporation into the cubic fcc lattice of gold. Results indicate that these nitrogen atoms prefer lattice tetrahedral sites as it happens for any TMN. Configurations 4OH–2TH and 6TH were the most energetically-stable structures for AuN when it has 60 % nitrogen atoms. However, for higher N concentrations, nitrogen atoms locate in each of tetrahedral sites, tending to generate the wurtzite hexagonal lattice. Finally, lattice parameters found are quite similar to the ones obtained experimentally, in addition, when the nitrogen is incorporated in the Au structure, relaxed parameters increase abruptly. Our results provide a consistent interpretation of the experimental data and could be used for understanding the growth of gold nitride. Therefore, from the results obtained in this research, it is expected that AuNx compound could be used in research aimed at the future development of new materials and the design of electronic devices.

ACKNOWLEDGMENTS

The authors gratefully acknowledge to the Office of the Vice-chancellor for Research at Universidad de Medellin and Colombian Department of Science, Technology and Innovation; in Spanish, Instituto Colombiano para el Desarrollo de la Ciencia y la Tecnología, COLCIENCIAS' funding provided through its program called "National Doctorates in the 2nd National Support Call for Graduate Programs in the 2006".

REFERENCES

[1] T. Maruyama, T. Morishita. Appl. Phys. Lett. 69 (1996) 890-891.
[2] J.H Quintero, R. Ospina, A. Mello. J of Phys: Conf Series 687 Issue 1 (2016) 012006.
[3] A. Zerr, G. Miehe, R. Boehler. Nat. Mater. 2 (2003) 185-189.
[4] N. Spyropoulos-Antonakakis, E. Sarantopoulou, Z. Kollia, G. Drazic, S. Kobe. Appl. Phys. Lett. 99 (2011) 153110.
[5] J.H. Quintero, R. Ospina, O.O. Cárdenas, G.I. Alzate and A. Devia. Phys. Scr. T131 (2008) 014013.
[6] Q. Zhan, R. Yu, L. He, D. Li, H. Nie, C. Ong. Mater. Lett. 57 (2003) 3904-3909.
[7] V. Ranjan, L. Bellaiche, E. J. Walter. Phys. Rev. Lett. 90 (2003) 257602.
[8] Q. Zhan, R. Yu, L. L. He, and D. X. Li. Thin Solid Films 411 (2002) 225-228.
[9] H. Al-Brithen, A. R. Smith. Appl. Phys. Lett. 77 (2000) 2485-2487.
[10] J. H. Quintero, P. J. Arango, R. Ospina, A. Mello and A. Mariño. Surf. Inter. Anal. 47 (2015) 701-705.
[11] S. Krishnamurthy, M. Montalti, M. G. Wardle, M. J. Shaw, P. R. Briddon, K. Svensson, M. R. C. Hunt, L. Siller. Phys. Rev. B 70 (2004) 045414.
[12] A.P. Caricato, M. Fernandez, G. Leggieri, A. Luches, M. Martino, F. Romano, T. Tunno, D. Valerini, A. Verdyan, Y.M. Soifer, J. Azoulay, L. Meda. Appl. Surf. Sci. 253 (2007) 8037-8040.
[13] A. Devia, H.A. Castillo, V.J. Benavides, Y.C. Arango, J.H. Quintero. Mater. Charac. 59 (2008) 105–107.
[14] E. S. Shanley, J. L. Ennis. Ind. Eng. Chem. Res. 30 (1991) 2503-2506.
[15] R. Yu, X. F. Zhang. Phys. Rev. B 72 (2005) 054103.
[16] R. Yu, X. F. Zhang. Appl. Phys. Lett. 86 (2005) 121913.
[17] M.B. Kanoun, S. Goumri-Said. Phys. Lett. A 362 (2007) 73-83.
[18] E. Zhao, J. Wang, J. Meng, Z. Wu. Comp. Mater. Sci. 47 (2010) 1064–1071.
[19] E. Zhao, Z. Wu. J. Solid State Chem. 181 (2008) 2814–2827.
[20] Mohammed S, H. Suleiman, and Daniel P. Joubert. [cond-mat.mtrl-sci] 15 sep 2013. arXiv: 1309.3753vl.
[21] L. Alves, T. P. A. Hase, M. R. C. Hunt, A. C. Brieva, L. Siller. J. Appl. Phys. 104 (2008) 113527.
[22] J.H Quintero, A. Mariño, P.J Arango. J Phys: Conf Series 466 (1) (2013) 012002.
[23] Evans, R. C., An Introduction to Crystal Chemistry, Cambridge Univ. Press, Cambridge (1979).
[24] J.P. Perdew, K. Burke, M. Ernzerhof. Phys. Rev. Lett. 77 (1996) 3865.
[25] K. Laasonen, A. Pasquarello, R. Car, C. Lee, D. Van-derbilt. Phys. Rev. B. 47 (1993) 10142.
[26] P. Giannozzi et al. J. Phys: Cond. Matt. 21 (2009) 395502.
[27] H. Monkhorst and J. Pack. Phys. Rev. B. 13 (1976) 5188.
[28] M. Methfessel and A. Paxon. Phys. Rev. B. 40 (1989) 3616.
[29] F. D. Murnaghan. Proc. Natl. Acad. Sci of USA 30 (1944) 30, 244-247.
[30] Hugh O. Pierson. Handbook of Refractory Carbides and Nitrides (Properties, Characteristics, Processing and Applications). Noyes Publications, 1996.
[31] A. Devia, V. Benavides, H. A. Castillo, J. Quintero. AIP Conf. Proc. 875 (2006) 258-261.
[32] J.H Quintero, A. Mariño, L. Šiller, E Restrepo-Parra, F.J. Caro-Lopera. Surf Coat Tec 309 (2017) 249-257.

Automatic aeroponic irrigation system based on Arduino's platform

A P Montoya, F A Obando, J G Morales and G Vargas

Universidad Nacional de Colombia, Medelln, Colombia. Calle 59A N63-20, Bloque 58

E-mail: apmontoy@unal.edu.co

Abstract. The recirculating hydroponic culture techniques, as aeroponics, has several advantages over traditional agriculture, aimed to improve the efficiently and environmental impact of agriculture. These techniques require continuous monitoring and automation for proper operation. In this work was developed an automatic monitored aeroponic-irrigation system based on the Arduino's free software platform. Analog and digital sensors for measuring the temperature, flow and level of a nutrient solution in a real greenhouse were implemented. In addition, the pH and electric conductivity of nutritive solutions are monitored using the Arduino's differential configuration. The sensor network, the acquisition and automation system are managed by two Arduinos modules in master-slave configuration, which communicate one each other wireless by Wi-Fi. Further, data are stored in micro SD memories and the information is loaded on a web page in real time. The developed device brings important agronomic information when is tested with an arugula culture (*Eruca sativa* Mill). The system also could be employ as an early warning system to prevent irrigation malfunctions.

1. Introduction

Aeroponics is a culture technique where the plant roots are suspended in the air and they are intermittently sprayed with a nutrient solution. This technique has been used both in research and crops commercial production [1, 2, 3, 4, 5]. Aeroponics presents great advantages over traditional agriculture, like reduced the consumption of water and nutrients, increased of growth rate, plant density and therefore crop productivity [6, 2, 7]. Its main disadvantage is crop susceptibility to electrical faults of the irrigation system [8, 7].

To optimize resources consumption and root's oxygenation [5], the nutrient solution is applied intermittently and recirculated, therefore it is necessary to continuously monitoring the consumption and level of the nutrient solution, as well as the correct operation of pumps and valves, because a system failure could cause crop loss, due to the lack of substrate of plants. It is also fundamental in this hydroponic systems monitoring the temperature, pH and electrical conductivity of the nutrient solution, since these variables are determinant in the nutrition of the crop.

Some work has been done on this regard. Reyes et al. [9], proposed the development of a monitoring and control system based on wireless communication protocol Xbee. In that article there is not a description of the materials and methodology used to develop the proposed irrigation system, which is based on the relative humidity within the root humidity-chamber. On the other hand, It has been some advances in the automation, monitoring and control of the nutritive solution in the production of potatoes seeds in aeroponic systems, providing roots with water and nutrients independently, using ultrasonic actuators for nutrients and sprinklers for water supply [10].

There are different automation solutions that can be used in aeroponic irrigation system, although its functionality is limited and costs and maintenance are high [11]. Designs that involves the use of open source software and hardware is an economical alternative for the development of automatic monitoring and control systems. Arduino is today one of the most popular open platforms used to obtain a large number of applications [12]. Recent applications in the agricultural area use Arduino as a platform for the development of control systems. Bitella et al. [13], Ferrarezi et al. [14], and Siva [15] worked in an automatic irrigation systems in soil cultivation. While Bajer & Krejcar [16], and Groener et al. [17], implemented a monitoring and control system for a greenhouse operation. Regarding aeroponics, Jonas et al. [18], developed an automatic monitoring system to control the environmental and irrigation conditions of a small aeroponic chamber using Arduino. The system controls the irrigation frequency based on the root chamber moisture. All the information is send to a web server and share in Twiter.

In the literature review it was not possible to find information related to the implementation of low cost systems for monitoring pH and electrical conductivity of hydroponics nutrient solution. There is a generalized lack of information in scientific literature related with the monitoring and automation of hydroponic irrigation systems.

This paper presents the construction of a low cost system for monitoring and automating the aeroponic irrigation of a greenhouse research facility. In addition to automate the irrigation frequency several variables associated with the nutrient solution and the proper functioning of the irrigation system are monitored, like the pH and electrical conductivity of the nutrient solution. The information provided by the system is stored in microSD memory and it can be accessed in real time by an Internet web server and Bluetooth.

2. Materials y methods

2.1. Irrigation system hardware

2.1.1. Greenhouse aeroponic system The implementation was carried out in a chapel-greenhouse facility with an area of 16x8m, average height 4m, built in steel structure and anti-drip thermal plastic gauge 6. The greenhouse has roller side windows activated by gearmotors. The facilities are located at 6.26, -75.57, altitude 1469 m.s.n.m, Medellin-Colombia.

The measurement of external climatic variables was performed with a weather station (Davis Vantage Pro 2). While the internal monitoring has been made using self-made aspiration thermo-higrometers.

The greenhouse has nine aeroponic-recirculating growing beds, with dimensions of 1x6m and maximum capacity of 2000 plants each. The beds were built in aluminum, black plastic and expanded polystyrene gauge 6 (EPS). Each bed has a hydraulic pump, nutrient solution tank and independent recirculation system, which included filters and 24 sprinkles (0.0097 $\frac{l}{s}$) each one. The diagram of the hydraulic system of two of the beds can be seen in Figure 1.

Due it was available only one pH sensor and one EC sensor, normally open and closed solenoid valves were used to make independently measurements of these variables for each bed.

2.1.2. Arduino's modules For monitoring and automating the irrigation of the aeroponic system, two Arduino Mega modules in master-slave configuration were used. They were packaged on PVC boxes with external connectors for instrumentation and communications. The master module is used for storing data in a microSD memory, it is also in charge of microclimatic data adquisition, Wi-Fi and Ethernet communications. The Arduino-slave was used for monitoring and automation of the irrigation system, it is also saved their own information in a microSD memory and is in charge of the bluetooth comunication. The devices used in the implementation can be appreciate in the connection diagram in Figure 2.

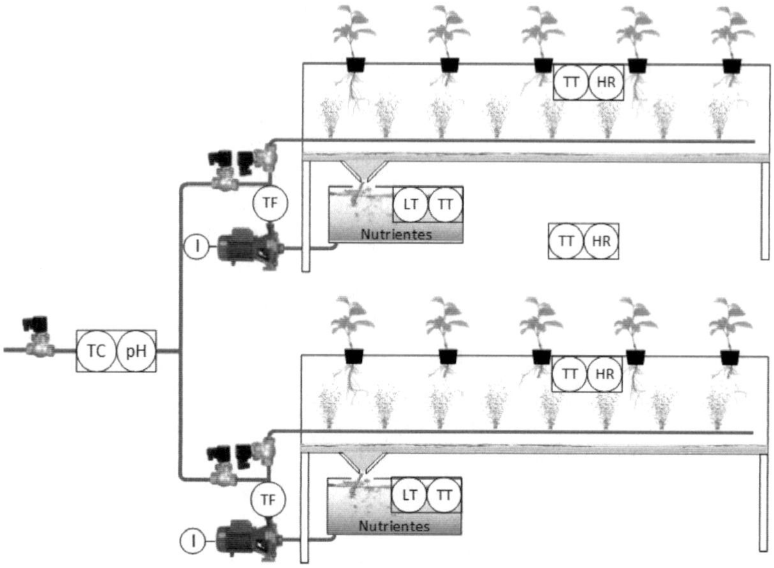

Figure 1. Implementation diagram.

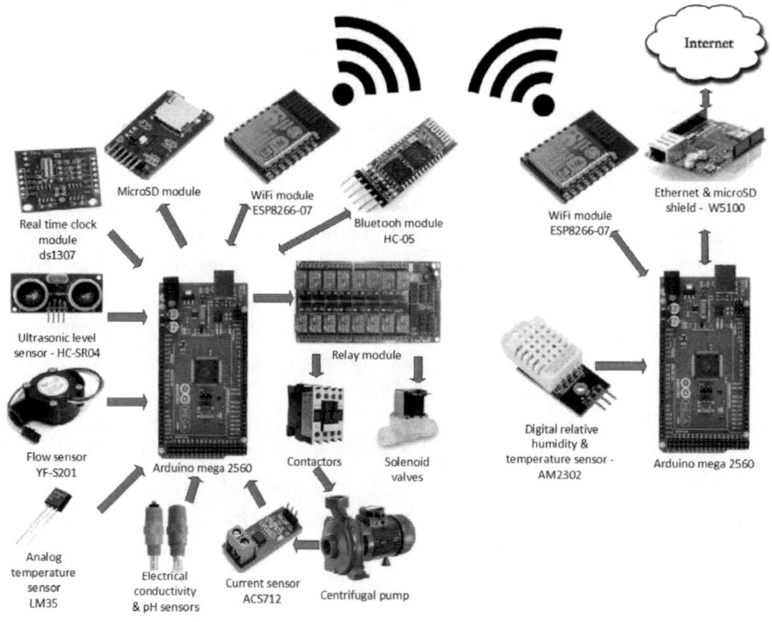

Figure 2. Detailed Master-Slave connection diagram.

Independent power sources of 12V and 5V, 6 and 4A respectively, were used to power all the sensors and solenoid valves. The hydraulic pumps were connected to 110 VAC.

Table 1 detailed the measured variables and sensors used in this implementation.

IOP Conf. Series: Journal of Physics: Conf. Series **850** (2017) 012003 doi:10.1088/1742-6596/850/1/012003

Table 1. Implemented sensors and measured variables.

Measure variable	Sensor	Sensor principle	Calibration curve
Nutritive solution temperature	LM35	Solid state	$T = 113.1v - 7.8593$
Level	HC-SR04	Ultrasound	$Level = 0.0003V^2 - 0.3058lt$
Flow	YF-S201	Hall effect	
Pump electric current	ACS-71	Hall effect	
Electrical conductivity	HI3001	Potentiometric	$EC = 4960.3v - 1359.1$
pH	HI1001	Electrolytic	$pH = 6.383v - 1.1702$

2.1.3. Instrumentation The nutritive solution temperature sensors were protected on aluminum-PVC probes, as shown in Figure 3. These probes were immersed into the storage bins, looking that the sensor was completely sealed. The probes were calibrated using analogous bulb thermometers as pattern. The resulting calibration equation can be seen in Table 1, were T is te temperature in Celsius degrees and v is the voltage output of the sensors, given in millivolts.

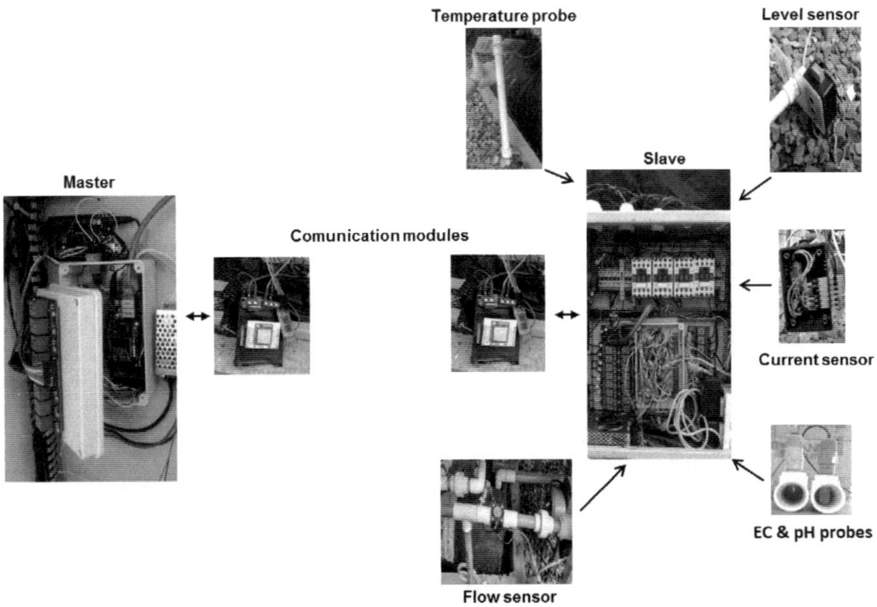

Figure 3. Implemented monitoring system.

The ultrasonic level sensors were installed on top of the storage bins. These were calibrated by measuring with a ruler and volumetric pattern of one liter. The resulting calibration equation is shown in Table 1, were V is the volume of nutritive solution given in liters and $Level$ is the actual level in cm.

The flow sensor was located in the pipe between the pump and the sprinklers. This sensor is pre-calibrated and the calibration curve was verified through timed volumetric measurement. The sensor was connected to the Arduino by one of the serial ports.

The electric current measurement is used to monitoring the proper functioning of the irrigation system. Through this variable is establishing the energy consumption of the hydraulic pump for each growing bed, verify the irrigation frequency and the proper functioning of the pumps. When the pump is working correctly the electric current oscillates around 2.42 - 2.18 A. When the current value is zero indicates that the pumps are off. If the current value exceeds the standard value, indicates that pump presents problems and should be revised.

The pH and EC sensors were installed in the common pipe of the growing beds. Such probes were connected to a HI-98143 transmitter, responsible for amplifying signals to 4-20 mA. This signal are adapted to voltage using precision resistors of 110 Ohm, to be acquired by the Arduino analogs inputs. The resulting voltage signal is differential, so it was necessary to configure four analog Arduino's inputs as differential inputs. To achieved this, Arduino's microcontroller registers were modified using AVR-C language, resulting in a library that allows analog data acquisition in differential mode with an adjustable gain between 10-200.

Since the measurement of pH and EC do not performed continuously and the probes most remained always submerged, the measurement routine included a process of pipe pre-cleaning with nutrient solution. After the measurement, the pipes and the three-way manifold, where sensors were installed, is washing and filling with clean water until the next measurement.

The probes were calibrated using pH and conductivity buffer solutions. The accuracy of the measurements was tested against a pH and EC meter using nutritive solutions at different concentrations. The obtained calibration equations can be seen in Table 1, where pH and EC are the values of pH and electrical conductivity (uS/cm^2) and v is the voltage value received in the differential pins of Arduino. The pH and EC electrodes were calibrated weekly to ensure proper operation and service. The measurements of these variables were temperature compensated.

2.1.4. Acting system This system includes four solenoid valves per growing bed, three valves are normally closed and one is normally open, as can be seen in Figure 4. In addition a hydraulic pump of $\frac{1}{2}$ HP, 36 $\frac{l}{min}$ flow and a maximum hydraulic head of 36 m is used. Besides, four electro-valves were used in the measurement and cleaning routine.

Figure 4. Aeroponic irrigation system actuators assembly.

All actuators were powered through a 16 channel relay module, which is connected to the digital outputs of slave Arduino. Plastic low pressure $\frac{1}{2}''$ diameter, 12 V - 0.5 A, solenoid valves were driven directly by the relay module, while the pumps were activated using additional contactors of 110 VAC. In Figures 1 and 4 it can be appreciate the detail of the devices that make up the system. Keeping in mind that not all elements are installed in the same place, for clarity the diagram of the process of the irrigation system is detailed in Figure 2.

2.1.5. Comunication Through bluetooth protocol may it access the information from the sensors and performing the test automation using the slave peripheral bluetooth module HC-05. This information can be accessed from any device with this kind of communication (phone, tablet, among others). Communication is established in the 2.4GHz band, with a range between 5 and 10 m away.

Master-Slave communication located 9m away one each other is done wirelessly via Wi-Fi. The wireless network is composed by two devices with password protected and has a range of 15-20m. The slave sends information related to the irrigation system. This data is stored redundantly in a 2GB microSD memory on the slave device, while the master store the same information besides the greenhouse climate data.

Theremore, the master device is connected to internet by Ethernet Industrial Protocol using the W5100 module. Information on climatic variables and irrigation is loaded to the website: `http://168.176.108.50:4025`, with a sampling rate of one minute. This information could be accessed from anywhere with Internet access. The Figure 5 shows a screenshot of the mentioned website, where it can be appreciate climatic and irrigation information in addition to alarm indicators.

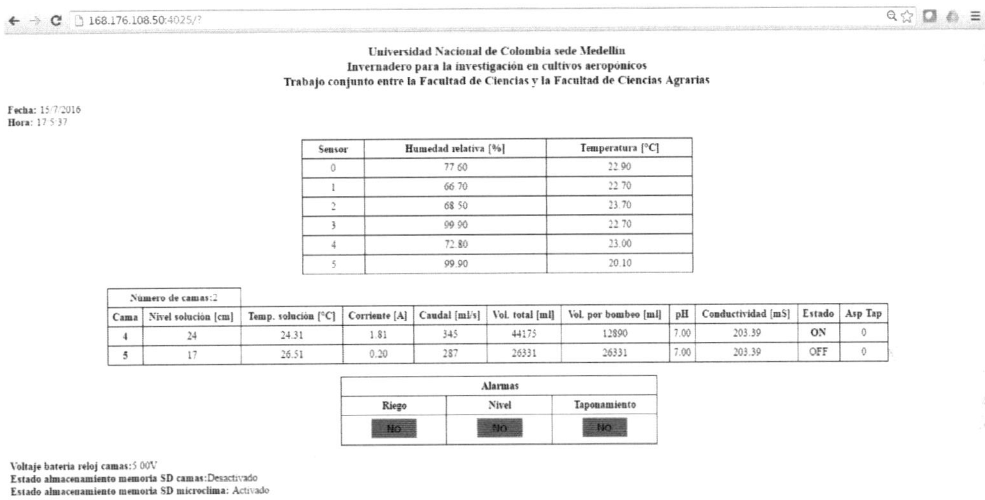
168.176.108.50:4025/?

Universidad Nacional de Colombia sede Medellín
Invernadero para la investigación en cultivos aeropónicos
Trabajo conjunto entre la Facultad de Ciencias y la Facultad de Ciencias Agrarias

Fecha: 15/7/2016
Hora: 17:5:37

Sensor	Humedad relativa [%]	Temperatura [°C]
0	77.60	22.90
1	66.70	22.70
2	68.50	23.70
3	99.90	22.70
4	72.80	23.00
5	99.90	20.10

Número de camas:2

Cama	Nivel solución [cm]	Temp. solución [°C]	Corriente [A]	Caudal [ml/s]	Vol. total [ml]	Vol. por bombeo [ml]	pH	Conductividad [mS]	Estado	Asp Tap
4	24	24.31	1.81	345	44175	12890	7.00	203.39	ON	0
5	17	26.51	0.20	287	26331	26331	7.00	203.39	OFF	0

Alarmas		
Riego	Nivel	Taponamiento
NO	NO	NO

Voltaje bateria reloj camas:5.00V
Estado almacenamiento memoria SD camas: Desactivado
Estado almacenamiento memoria SD microclima: Activado

Figure 5. Web page screenshot.

2.2. Arduino's software

In the master module was programmed routines for climate data acquisition, wireless communication, microSD memory storage and publication of data on the web. Whereas, in the slave device was programmed the same routines, except the web page publication one. Additional routines for irrigation and measuring EC and pH were programmed. The flow diagram of the main routine of both modules can be seen in Figure 6.

In the slave module was programmed diurnal and nocturnal irrigation frequency for each growing bed. The irrigation frequency is synchronized with the data acquisition , so when an irrigation cycle is performed the data are save in the microSD and send to the master module along with the acquisition time given by the Real Time Clock module.

They were also scheduled three alarms to alert system malfunction. The Alarm of the nutrient solution ending is activated when the liquid level in the storage tank reaches a minimum value (5 cm). The Alarm for the clogged sprinklers is activated when there is more than 4 sprinklers

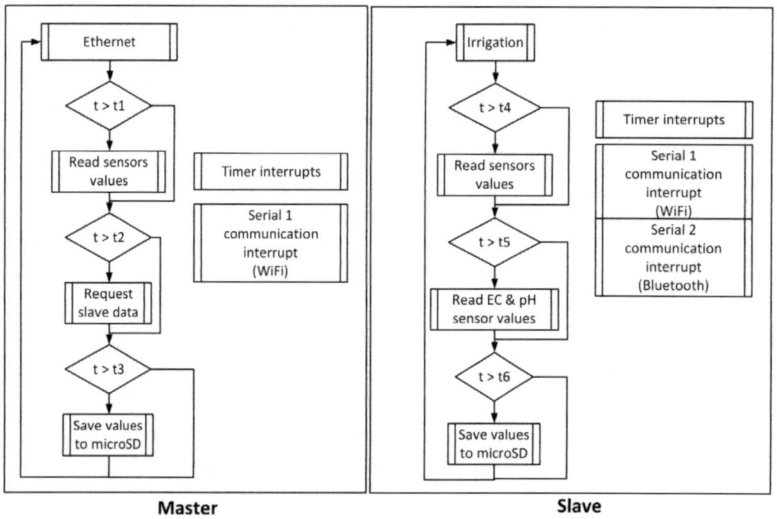

Figure 6. Main program flow diagram of master and slave modules.

plugged on a bed, an then the flow supplied by the pump is below minimum value. Finally, the alarm for the shutdown system is activated when a pump is activated and the pumped flow is zero. Indirectly it is possible to know electricity failures or Internet connection malfunction, if a pump is not active for long time or when the information on the website is not updated every minute. In Figure 5, it can be seen a detail of these alarms.

2.3. System test

An arugula crop (*Eruca sativa* Mill) was used to tests the implemented irrigation system. It was stablished a seedbed with 300 arugula seeds, planted in individual plastic containers with pine sawdust substrate. The plants were transplanted to the aeroponic beds at cotyledon state, when it was observed that the root protruded of the container (five days after seeding). System performance was recorded for 60 days between the months of May to July 2016.

The tests were performed on two beds, one with the arugula crop and the other one totally seal without crop. The irrigation frequency was established in 15 seconds turned on by 400 seconds turned off, using a modified Hoagland nutrient solution, diluted 50% with EC between 1200-1500 $\frac{uS}{cm^2}$ and pH between 5.5-6.5, measured at the preparation time. The diluted Hoagland nutrient solution was change one week after transplant for a full one, with EC and pH between 5.5-6.5 and 2100-2500 $\frac{uS}{cm^2}$, respectively, measured at preparation time. The nutrient solutions were manually prepared for each bed. The tanks were filled with 90 L of nutrient solution, equivalent to 31 cm in their respective bins. Independent tests were performed to monitor the consumption of the solution and the evolution of EC, pH, temperature solution, flow and pump electrical current in both beds.

3. Results y discussion

The external and internal climatic conditions during the test are show in Figure 7. The maximum and minimum internal temperature were 44 °C and 17 °C respectively, while the relative humidity was between 9-100 %.

Figure 8 shows a the nutrient solution consumption in the two beds during 78 hours of continuous operation. The identical tanks were filled with 87 L of nutritive solution. The minimum level was set at 27 L,to prevent plants lost. Both growing beds were totally sealed

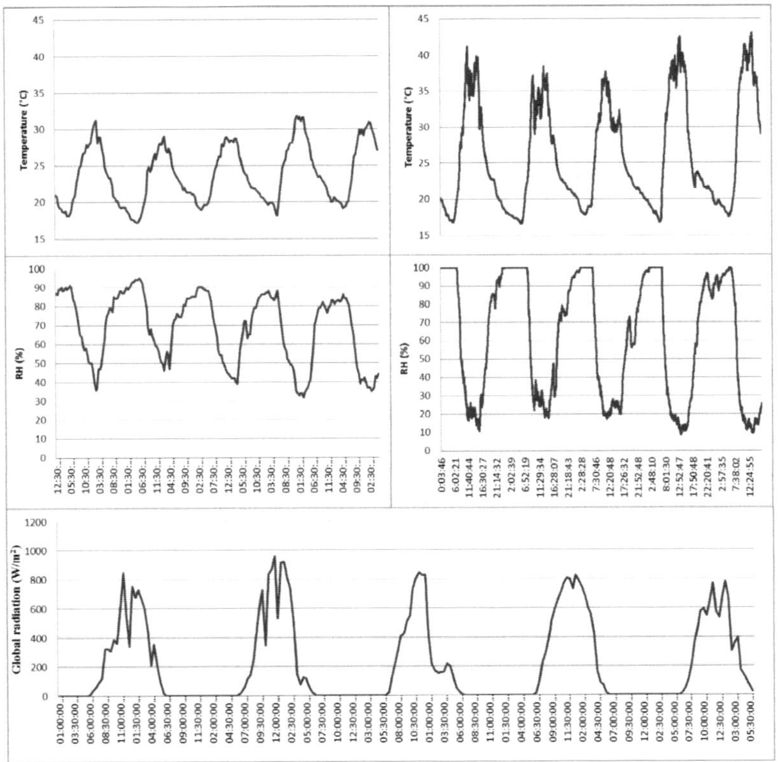

Figure 7. Climatic condition during five test days.

except for the holes where the plants are located. The bed without plants decreased their nutrient solution level due to evaporation, while in the growing bed with crop, the consumption of the solution is due to evapotranspiration. Subtraction of both consumptions results in crop transpiration for each time instant.

The total evaporation during the 78 h test was 11.91 L, while the evapotranspiration was 63.12 L, resulting in a crop transpiration of 51.07 L during the 78 h. The resulting individual plant transpiration was 2.18 $\frac{mL}{h}$. As expected, the greatest changes in evaporation occur during the day, which have higher temperatures and lower humidities. To avoid interfering with the nutrient solution consumption monitoring trials, independent tests for monitoring EC, pH, temperature and flow were made.

In Figure 9, at 9:30am of the first day, it can be see the change of the old Hoagland nutritive solution for a new one. The old solution (half Hoagland) were replaced with 90 L of a complete Hoagland solution for the bed with crop (pH 6). The other bed is running with tap water (pH 7).

It can be seen that the first measurement after the solution change, has a pH of 6.3, this value is due to the traces of the old nutritive solution in the pipes, so both pH values are slightly above of the the actual value at the time of the preparation. Small variations in pH are seen during the test, showing slight increases during the day and a tendency to increase the pH of the solution as it is consumed.

Regarding the electrical conductivity of the nutritive solution, in the Figure 10 is possible to appreciate the moment of the solution change, from a diluted Hoagland solution (50%), with conductivity 1400 $\frac{uS}{cm^2}$, for the complete solution (100%) with 2500 $\frac{uS}{cm^2}$. Due to the

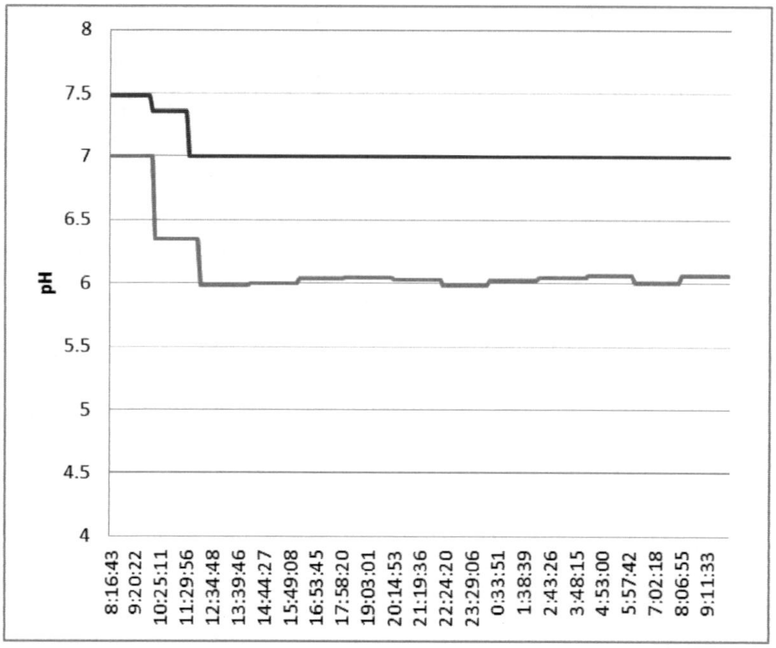

Figure 8. Nutritive solution consumption comparison during 78 h. For a growing bed without plants (blue) and a bed with 300 plants (red). The green line represents the crop transpiration.

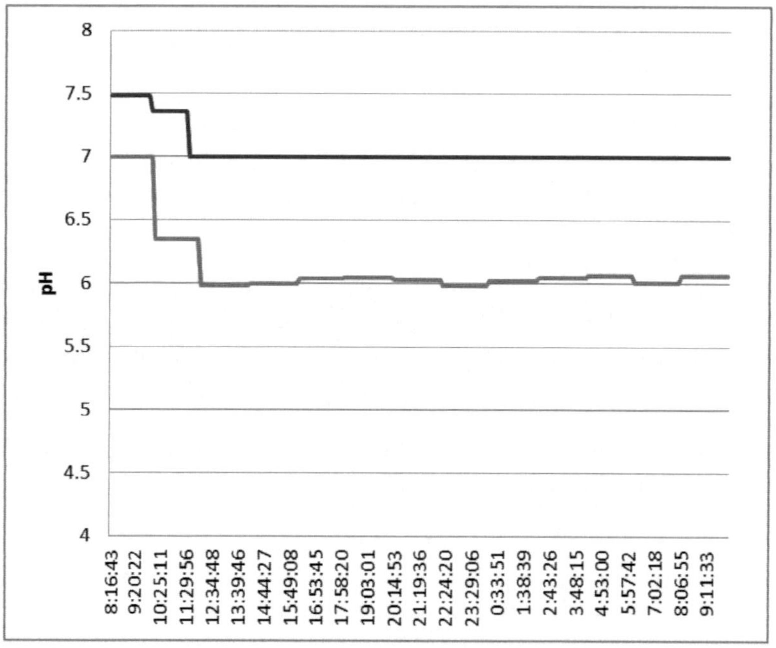

Figure 9. Nutritive solution pH comparison for a growing bed without plants (blue) and a bed with 300 plants (red).

instrument accuracy (200 $\frac{uS}{cm^2}$), it was not possible to conclude about the variations of that variable throughout the day, but it is possible to establish that there is a tendency to increase the conductivity of the nutrient solution during the test in the crop bed.

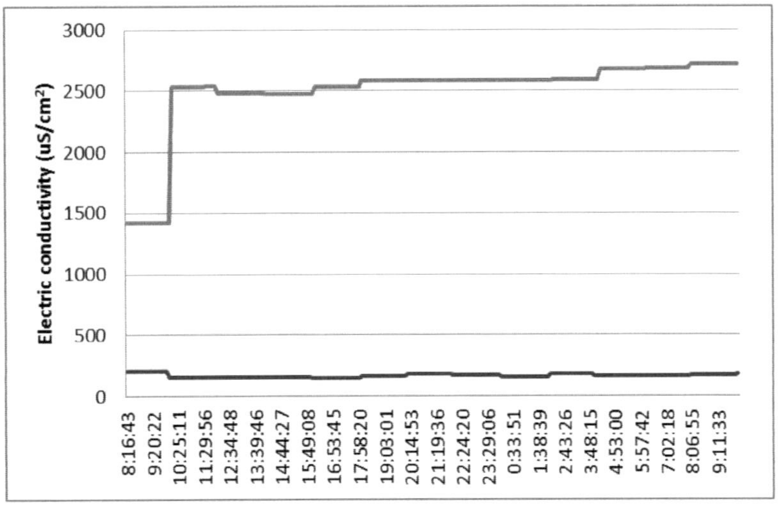

Figure 10. Nutritive solution EC comparison for a growing bed without plants (blue) and a bed with 300 plants (red).

The results of temperature measurements are presented in Figure 11. It can be appreciated that the temperature of the nutritive solution of the bed with crop has a mean temperature of 25 $^{\circ}C$, while the bed without crop has a mean of 24 $^{\circ}C$. The LM35 temperature sensor has an accuracy of $\pm 1\,^{\circ}C$, so that it is possible to consider that both values are the same and the temperature of the solution remains relatively constant throughout the experiment.

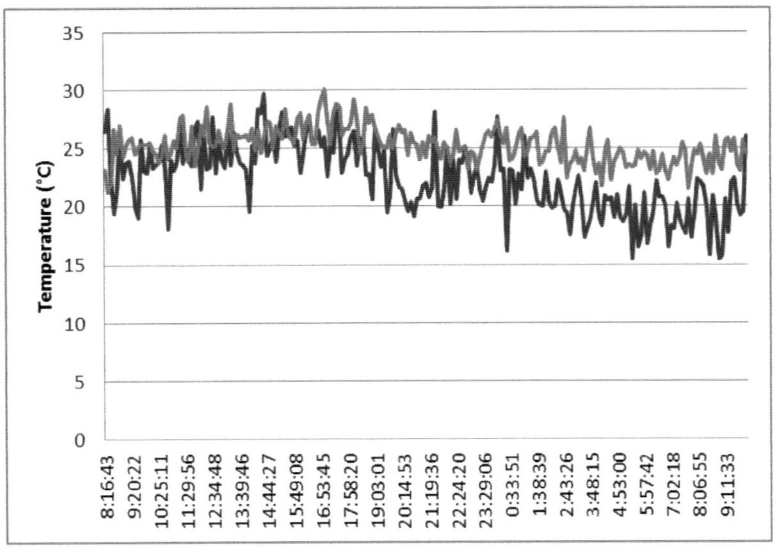

Figure 11. Nutritive temperature solution comparison for a growing bed without plants (blue) and a bed with 300 plants (red).

4. Conclusions

The automatic monitored aeroponic-irrigation system based on the Arduino free software platform present good performance when it is implemented in a research greenhouse. This results can be applied in a commercial greenhouse with polyculture or when different nutrient solutions formulation are employed. For this particular application the most difficult implementation task lies in the fact of having just one EC and pH probes for monitoring different nutrient solutions. In a commercial application, such difficulty is eliminated, since these systems only has a central tank of nutrient solution, therefore, the measurements can be realized continuously.

Arduino platform can be used as a data acquisition system for differential output sensors like EC and pH probes, using the differential configuration showed here, reducing the cost of such implementations.

The deployment of information and alarms on a website is very useful, allowing to take appropriate control irrigation actions to avoid crop loss.

It is considered relevant for future work automatically send alarm signals directly to the final user of the system through mobile devices. In addition, there is a desirable to develop a complementary automatic system to prepare the nutrient solution based on the EC and pH.

References

[1] Komosa A, Kleiber T and Markiewicz B 2014 *Acta Sci. Pol., Hortorum Cultus* **13** 163–177 ISSN 1644-0692
[2] Lakkireddy K, Kasturi K and Sambasiva Rao K R S 2012 *Research & Reviews: ...* **1** 26–35 ISSN 2278 2206 URL `http://stmjournals.com/sci/index.php?journal=RRJoAST{\&}page=article{\&}op=view{\&}path[]=50`
[3] Mbiyul M W, Muthoni J, Kabiral J, Elmar G, Muchiral C, Pwaipwail P, Ngaruiyal J, Otienol S and Onditil J 2012 *Journal of Horticulture and Forestry* **4** 172–177 URL `http://www.academicjournals.org/journal/JHF/article-abstract/C5947BF2821`
[4] Nir I 1981 Growing Plants in Aeroponics Growth System
[5] Peterson L A and Krueger A R 1988 *Crop Science* **28** 712–713 ISSN 1098-6596 (*Preprint* `arXiv:1011.1669v3`)
[6] Chiipanthenga M, Maliro M, Demo P and Njoloma J 2012 *African Journal of Biotechnology* **11** 3993–3999
[7] Putra P A and Yuliando H 2015 *Agriculture and Agricultural Science Procedia* **3** 283–288 ISSN 22107843 URL `http://www.sciencedirect.com/science/article/pii/S2210784315000558`
[8] Carruthers S 1992 *Practical Hydroponics & Greenhouses*
[9] Reyes J L, Montoya R, Ledesma C and Ramírez R 2012 *Acta Horticulturae* **947** 153–156 ISSN 05677572
[10] Idris I and Sani M I 2012 *Proceedings of 2012 IEEE Conference on Control, Systems and Industrial Informatics, ICCSII 2012* 120–125
[11] Gaddam A, Al-Hrooby M and Esmael W F 2014 Designing a Wireless Sensors Network for Monitoring and Predicting Droughts *Proceedings of the 8th International Conference on Sensing Technology* pp 210–215 ISSN 21568073
[12] Mesas-Carrascosa F, Verdú Santano D, Meroño J, Sánchez de la Orden M and García-Ferrer A 2015 *Biosystems Engineering* **137** 73–83 ISSN 15375110 URL `http://www.sciencedirect.com/science/article/pii/S1537511015001208`
[13] Bitella G, Rossi R, Bochicchio R, Perniola M and Amato M 2014 *Sensors (Switzerland)* **14** 19639–19659 ISSN 14248220
[14] Ferrarezi R S, Dove S and Van Iersel M W 2014 A Low Cost, Arduino-based System for Monitoring and Controlling Substrate Water Content *2014 ASHS Annual Conference*
[15] Siva Kumar P, Jilani S A K and Venkata Raman B 2015 *International Journal of Advanced Research in Electronics and Communication Engineering (IJARECE)* **4**
[16] Bajer L and Krejcar O 2015 *IFAC-PapersOnLine* **48** 368–373 ISSN 24058963 URL `http://www.sciencedirect.com/science/article/pii/S240589631500837X`
[17] Groener B, Knopp N, Korgan K, Perry R, Romero J, Smith K, Stainback A, Strzelczyk A and Henriques J 2015 *Procedia Engineering* **107** 470–479 ISSN 18777058 URL `http://dx.doi.org/10.1016/j.proeng.2015.06.105`
[18] Jonas P, Maskara A, Salguero A and Truong A 2015 Garduino: A Cyber-Physical Aeroponics System (*Preprint* `arXiv:1011.1669v3`) URL `http://ecal.berkeley.edu/files/ce186/projects/truonganders{\_}4941954{\_}63829747{\_}Garduino-1.pdf`

Dynamic analysis and control PID path of a model type gantry crane

P. A. Ospina-Henao[1,2], Framsol López-Suspes[1,2]

1. Grupo de Investigación en Ciencias Básicas y Aplicadas. Departamento de Ciencias Básicas, Universidad Santo Tomás, Carrera 18 No. 9 - 27 PBX 6800801, Bucaramanga, Colombia.
2. Grupo de Investigación en Relatividad y Gravitación, Escuela de Física, Universidad Industrial de Santander, A. A. 678, Bucaramanga, Colombia.

E-mail: paospina@uis.edu.co, framsol.lopez@ustabuca.edu.co

Abstract. This paper presents an alternate form for the dynamic modelling of a mechanical system that simulates in real life a gantry crane type, using Euler's classical mechanics and Lagrange formalism, which allows find the equations of motion that our model describe. Moreover, it has a basic model design system using the SolidWorks software, based on the material and dimensions of the model provides some physical variables necessary for modelling. In order to verify the theoretical results obtained, a contrast was made between solutions obtained by simulation in SimMechanics-Matlab and Euler-Lagrange equations system, has been solved through Matlab libraries for solving equation's systems of the type and order obtained. The force is determined, but not as exerted by the spring, as this will be the control variable. The objective is to bring the mass of the pendulum from one point to another with a specified distance without the oscillation from it, so that, the answer is overdamped. This article includes an analysis of PID control in which the equations of motion of Euler-Lagrange are rewritten in the state space, once there, they were implemented in Simulink to get the natural response of the system to a step input in F and then draw the desired trajectories.

1. Introduction

There are many practical problems that require control system to be solved properly. Many of these problems involve highly complex dynamic systems, which often include nonlinearities, but can be approximated by more simple representations, i.e. the gantry type container cranes, used for the loading and unloading of ships in ports. The mass of the container, suspended from a trolley wire that carries along a rail, it presents a similar dynamic to the problem of the carriage and pendulum. This problem was configured as a case of an underactuated system: is necessary to control two variables, the carriage position and the mass by a single input signal, the force applied to the carriage. In addition, there are strict performance requirements: high response speed (cannot waste time in loading and unloading operations) and absence of oscillations in the position of the mass (the back and forth motion of the container can be dangerous). Container cranes are formidable machines typically employed in port transhipment hubs for cargo movement whose efficiency, directly linked to the crane effective speed, strongly influences the productivity of the overall terminal. To optimize the crane performance in terms of minimization of the working operation time, the accurate characterization of the dynamic response of containers during the lifting/lowering maneuvering phase is an essential requirement.

Moreover, harbors are often located in windy areas, characterized by the occurrence of gust fronts that can affect the cranes working efficiency and, sometimes, compromise the safety of workers.

In the literature, several studies on container cranes dynamics are available and different modeling approaches (*e.g.*, planar or three-dimensional models with elastic/rigid hoisting cables) are proposed to investigate their behavior. A comprehensive classification of these models is given in [1] where applications and limitations of the different theories are discussed. In order to investigate the dynamic behavior of container cranes, in recent years several models have been developed and control strategies have been proposed to overcome the potential issues arising from uncontrolled large-amplitude oscillations. Studies of container crane pendulations were first conducted in [2] where the dynamic behavior of simply suspended objects was investigated and an extensive review of the theory was presented. In the same work, a control strategy, based on programmed acceleration/deceleration phases of the trolley motion, was proposed and numerical results were compared with experiments performed on a scale model. Payload oscillations may also be excited, as in ship-mounted cranes, from the sea wave motion that can induce oscillations of the ship and, consequently, of the suspended container. Wave induced oscillations were described in [3] where the planar motion of a one degree of freedom container crane model was studied and a delayed position feedback controller was proposed.

Numerical simulations, aimed to study the effects of non linearities arising from mechanical friction and air resistance, were performed in [4] where the results were also experimentally validated. In [5]-[6], modeling of a variety of pendular structures (such as gantry cranes) was proposed and suitable control strategies were thoroughly examined by highlighting the computational implications of the models on the control implementations. A 1/10 laboratory scale model of a gantry crane was employed in [5] to test the performance of a feedback linearized control method, whereas in [6] single- and multi-cable models of gantry cranes were discussed.

The main purpose of controlling an underactuated crane system is transporting the payload in a precise location. However, it is very difficult due to the fact that the payload can exhibit a pendulum-like swinging motion. Various attempts in controlling cranes system based on open loop and closed-loop control system have been proposed. For example, open loop time optimal strategies were applied to the crane by many researchers [7]-[8]. For example, PD controllers has been proposed for both position and anti-swing controls [9]-[10]. However, the performance of the controller is not very effective in eliminating the steady state error.

The dynamics of the car-pendulum type gantry crane is more complex than the single pendulum-type overhead crane, so the system analysis and control algorithm design also is more difficult. In this paper, the gantry crane that exhibits car-pendulum dynamics is investigated by using the Euler-Lagrange equations [11]-[12]. The system model is built and the system properties are analyzed. An adaptive PID control algorithm is proposed for the car-pendulum type gantry crane and the simulation results demonstrate the system dynamics and the effectiveness of the proposed control algorithm. The PID controller is the most common form of feedback. It was an essential element of early governors and it became the standard tool when process control emerged in the 1940s. In process control today, more than 95% of the control loops are of PID type, most loops are actually PI control. PID controllers are today found in all areas where control is used. The controllers come in many different forms. There are stand alone systems in boxes for one or a few loops, which are manufactured by the hundred thousands yearly. PID control is an important ingredient of a distributed control system. The controllers are also embedded in many special purpose control systems [13]. PID control is often combined with logic, sequential functions, selectors, and simple function blocks to build the complicated automation systems used for energy production, transportation, and manufacturing. Many sophisticated control strategies, such as model predictive control, are also organized hierarchically. PID control is used at the lowest level; the multivariable controller gives the setpoints to the controllers at the lower level.

The remainder of this paper is organized as follows. The dynamics of an actual configuration of a gantry crane transporting distributed mass payloads, and we obtain the Euler-Lagrange equations for the model, is presented in Section 2. Next, in section 3, we presented the numerical model from based in the Solidworks and the comparative graphs in the analytical and the numerical solutions. In the Section 4, we presented the theory of small oscillations, we verify the natural frequencies and normal coordinates is validate. In the Section 5, the PID control of the system is described above and experiments on a gantry crane and the controller design procedure is presented. Block diagram of the system and the two variables which require control we presented. Finally, in Section 6, we present the conclusions of our main results.

2. Symbolic model

The crane type modeling system started from a model in which intervenes a spring at the point where the force is applied for the movement of the carriage on the rail, as shown in Figure 1 and the real gantry crane as shown in the Figure 2.

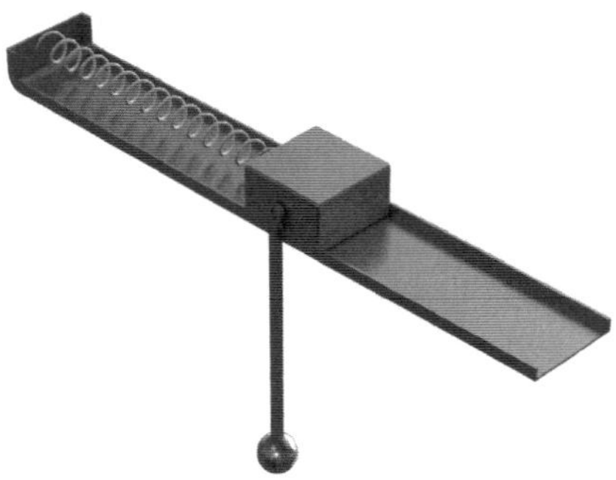

Figure 1. Initial system image

Figure 2. Gantry crane

For this system the Lagrangian model was developed, in which initially the kinetic and potential energies are calculated. For this model, was defined as M is the mass of the carriage (object on the rail), m_1 the mass of the extreme of the pendulum, m_2 the mass of the pendulum rod, θ the angle between the vertical and the pendulum, the pendulum, l the length and, x the displacement of the carriage.

2.1. Kinetic energy

The kinetic energy is defined

$$T = \frac{1}{2} m v^2, \tag{1}$$

where m is the mass, v velocity, and T the kinetic energy of the body. In this case there are two bodies, the cart and the pendulum. The kinetic energy of the cart is,

$$T_1 = \frac{1}{2} M \dot{x}^2, \tag{2}$$

the translational kinetic energy of the extreme of the pendulum,

$$T_2 = \frac{1}{2} m_1 \left(l^2 \dot{\theta}^2 + 2 l \dot{\theta} \dot{x} \cos(\theta) + \dot{x}^2 \right), \tag{3}$$

the translational kinetic energy of the pendulum rod,

$$T_3 = \frac{1}{2} m_2 \left(\frac{1}{4} l^2 \dot{\theta}^2 + l \dot{\theta} \dot{x} \cos(\theta) + \dot{x}^2 \right), \tag{4}$$

IOP Conf. Series: Journal of Physics: Conf. Series **850** (2017) 012004 doi:10.1088/1742-6596/850/1/012004

the rotational energy of the pendulum rod is,

$$T_4 = \frac{1}{24} m_2 \dot{\theta}^2 l^2, \tag{5}$$

and the rotational energy of the extreme of the pendulum is,

$$T_5 = \frac{1}{2} m_1 \dot{\theta}^2 \left(\frac{2}{5} r^2 \right), \tag{6}$$

where, r is the radius of the sphere (extreme of the pendulum).

2.2. Potential energy

This system stores potential energy in the spring and the pendulum, for spring,

$$U_1 = \frac{1}{2} K x^2, \tag{7}$$

where, K is the elasticity constant of the spring. The potential energy of the pendulum is,

$$U_2 = -g\, l \cos(\theta) \left(m_1 + \frac{m_2}{2} \right), \tag{8}$$

where, g represents the value of the gravitational acceleration.

2.3. Euler-Lagrange equations

The Lagrangian of any system is defined,

$$\mathcal{L} = T - U, \tag{9}$$

where, T is the kinetic energy and U the potential, therefore, for this model it is,

$$\begin{aligned} \mathcal{L} &= T_1 + T_2 + T_3 + T_4 + T_5 - U_1 - U_2 \\ &= \tfrac{1}{2} M \dot{x}^2 + \tfrac{1}{2} m_2 \left(\tfrac{1}{4} l^2 \dot{\theta}^2 + l \dot{\theta} \dot{x} \cos(\theta) + \dot{x}^2 \right) \\ &+ \tfrac{1}{2} m_1 \left(l^2 \dot{\theta}^2 + 2 l \dot{\theta} \dot{x} \cos(\theta) + \dot{x}^2 \right) \\ &+ \tfrac{1}{24} m_2 l^2 \dot{\theta}^2 + \tfrac{1}{2} m_1 \dot{\theta}^2 \left(\tfrac{2}{5} r^2 \right) \\ &+ g\, l \cos(\theta) \left(m_1 + \tfrac{m_2}{2} \right) - \tfrac{1}{2} K x^2. \end{aligned} \tag{10}$$

The Euler-Lagrange equations are given by the following equation

$$\frac{d}{dt} \left(\frac{\partial \mathcal{L}}{\partial \dot{q}} \right) - \frac{\partial \mathcal{L}}{\partial q} = 0, \tag{11}$$

where, q represents the degrees of freedom (DOF) generalized, in this case, there are two DOF, θ and x, therefore, from equation (11),

$$\frac{d}{dt} \left(\frac{\partial \mathcal{L}}{\partial \dot{\theta}} \right) - \frac{\partial \mathcal{L}}{\partial \theta} = 0 \tag{12}$$

$$\frac{d}{dt} \left(\frac{\partial \mathcal{L}}{\partial \dot{x}} \right) - \frac{\partial \mathcal{L}}{\partial x} = 0. \tag{13}$$

From (12) and (13), the Euler-Lagrange equation, is obtained,

$$\ddot{\theta}l^2\,A + [\ddot{x}\,l\cos(\theta) + g\,l\sin(\theta)]\;B = 0 \tag{14}$$

$$\ddot{x}C - \left[\ddot{\theta}l\cos(\theta) - l\dot{\theta}^2\sin(\theta)\right]B + K\,x = 0, \tag{15}$$

where,

$$A = \left(m_1 + \frac{m_2}{3}\right) + \frac{2\,m_1\,r^2}{5} \tag{16}$$

$$B = m_1 + \frac{m_2}{2} \tag{17}$$

$$C = m_1 + m_2 + M \tag{18}$$

3. Numerical model

To define the numerical model, the constant values are replaced,

Table 1. Constant values in the model.

Constant	Value	
l	=	0.2 [m]
g	=	9.81 [m/s^2]
K	=	100 [N/m]
m_1	=	0.088338025 [Kg]
m_2	=	0.022245336 [Kg]
M	=	0.548069759 [Kg]
r	=	0.02 [m]

Substituting these values in equations (14) and (15),

$$0.003844\,\ddot{\theta} + 0.19514\sin(\theta) + 0.01989\,\ddot{x}\cos(\theta) = 0 \tag{19}$$

$$0.6586\,\ddot{x} - 0.01989\,\dot{\theta}^2\sin(\theta) + 0.01989\,\ddot{\theta}\cos(\theta) + 100\,x = 0 \tag{20}$$

As can be seen, the equations (19) and (20) are coupled, that is to say, both depend on two variables to their second derivative, $\ddot{\theta}$, $\ddot{x}$, therefore, the initial conditions depend on the initial energy that shows this system, then, to find the initial conditions that can take, part of the potential energy is taken, which it is the energy that accumulates the spring and the pendulum when θ is different from zero. The total potential energy of the model is

$$\begin{aligned} U_T &= U_1 + U_2 \\ &= -g\,l\cos(\theta)\left(m_1 + \tfrac{m_2}{2}\right) + \tfrac{1}{2}K\,x^2, \end{aligned} \tag{21}$$

replacing the numerical values of the constants,

$$U_T = -0.19514\cos(\theta) + 50\,x^2. \tag{22}$$

There are infinite energy values that could take the system, however, only interested in those who do not turn the model into a chaotic system, since it is assuming (until this moment) from conservative way, i.e., that no gain or loss of energy and therefore the energy level that present, is due to the initial position of the pendulum and spring's compression. For spring it can be assumed values between -0.1 and 0.1, and for the angular position of the pendulum $-\pi$ and π. From these values can be graphed contours for energy values 0, 0.1 and 0.2, considering that the reference system (zero potential energy) is located in the center of rotation of the pendulum. In Figure 3, the potential energy levels for the system is shown, and in Figure 4 the contours for three different energy levels.

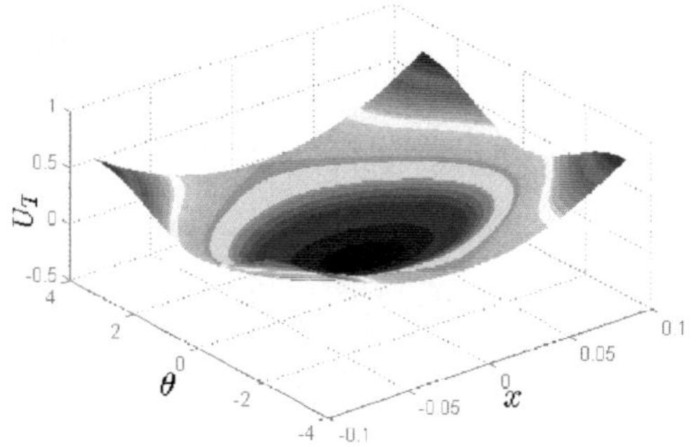

Figure 3. Energy level graph for the system

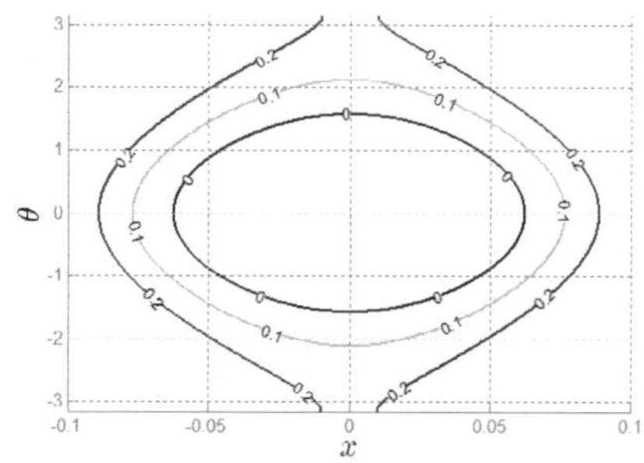

Figure 4. Contours for energy levels 0, 0.1 and 0.2

Now, you can take the zero energy level, because it is stable curve is completely closed (as the graph in Figure 3), so that this curve the values of initial conditions are taken to solve the coupled differential equations (19) and (20). For convenience and ease, a value is taken as zero, $\theta = 0$, for which $x = 0.0624$. So that the initial conditions are

Table 2. Initial conditions

Conditions	Value	
$x(0)$	=	0.0624
$\dot{x}(0)$	=	0
$\theta(0)$	=	0
$\dot{\theta}(0)$	=	0

The numerical solution was performed in MATLAB, entering the equations (19) and (20), followed by the next commands

```
V = odeToVectorField(eq1, eq2);
M = matlabFunction(V, 'vars', {'t', 'Y'});
sol = ode45(M, [0 10], [0.0624 0 0 0]);
```

where `eq1, eq2`, represent the numerical equations, and `sol`, the system solution, in this case MATLAB returns a structure (`struct`), in which are contained a time vector and a matrix of four rows, each of these is a solution of the variables, which order is $x, \dot{x}, \theta$ and $\dot{\theta}$ respectively. The solution for θ is shown in Figure 5 and x for in Figure 6.

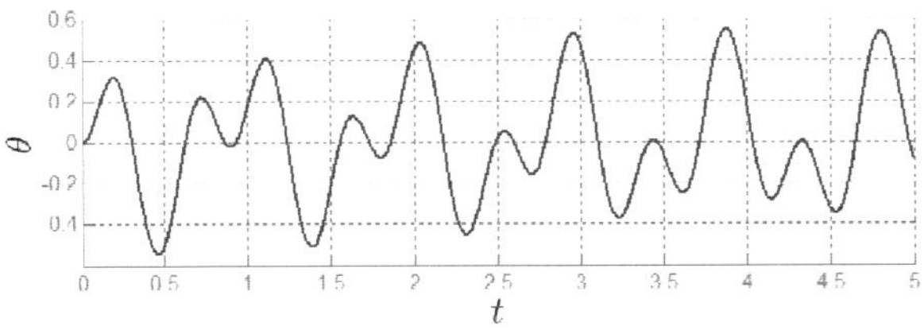

Figure 5. θ solution for the first 5 sec.

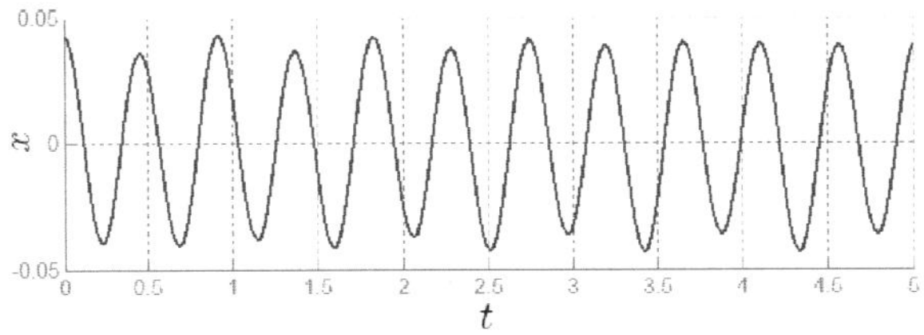

Figure 6. x solution for the first 5 sec.

To validate the results obtained, the system was performed in SolidWorks and their movement was simulated considering the initial conditions, when comparing the results obtained analytically with the simulation of SolidWorks, a great approach is observed, so you can have certainty about calculations performed from the Lagrangian's dynamics. Comparative graph is shown in Figure 7.

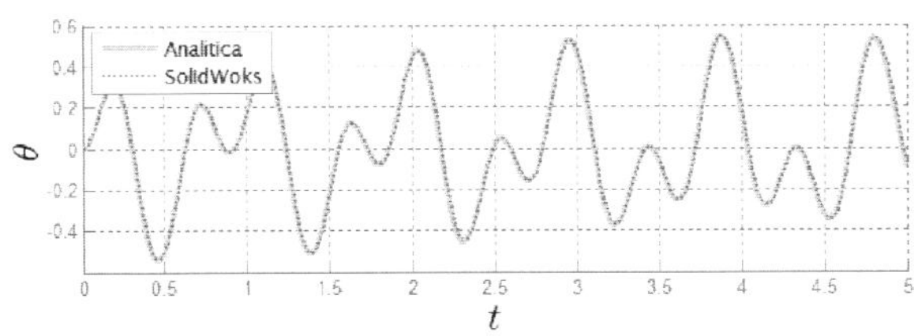

Figure 7. Comparative graph between analytical solution and simulation in SolidWorks.

4. Oscillations

The theory of small oscillations finds widespread physical applications in acoustics, molecular spectra, vibrations of mechanisms, and coupled electrical circuits. If the deviations of the system from stable equilibrium conditions are

small enough, the motion can generally be described as that of a system of coupled linear harmonic oscillators.

With the approximations for small angles

$$\sin(\theta) \approx \theta; \quad \cos(\theta) \approx 1; \quad \theta\,\dot{\theta}^2 \approx 0, \tag{23}$$

the kinetic and potential energy are

$$T = \frac{1}{2}\dot{\theta}^2\left[m_1\,l^2 + \frac{7}{12}m_2\,l^2\right] + \frac{1}{2}\dot{x}^2\left[m_2 + M\right] + \dot{\theta}\,\dot{x}\,l\left[m_1 + \frac{1}{2}m_2\right] \tag{24}$$

$$U = \frac{1}{2}K\,x^2 + \frac{1}{2}\theta^2 g\,l\left(m_1 + \frac{1}{2}m_2\right). \tag{25}$$

It is obvious from their definition that the U are symmetrical, that is, that $U_{ij} = U_{ji}$. A similar series expansion can be obtained for the kinetic energy. Since the generalized coordinates do not involve the time explicitly, the kinetic energy is a homogeneous quadratic function of the velocities. Expressing equations (24) and (25) matrix form

$$\mathbf{T} = \begin{bmatrix} l^2\left(m_1 + \frac{7}{12}m_2\right) & l\left(m_1 + \frac{1}{2}m_2\right) \\ l\left(m_1 + \frac{1}{2}m_2\right) & m_2 + M \end{bmatrix} \tag{26}$$

$$\mathbf{U} = \begin{bmatrix} g\,l\left(m_1 + \frac{1}{2}m_2\right) & 0 \\ 0 & K \end{bmatrix} \tag{27}$$

where the matrix (26) and (27) meet the following two equations

$$\mathbf{T} = \frac{1}{2}\mathbf{T}_{ij}\,\dot{\theta}\,\dot{x} \tag{28}$$

$$\mathbf{U} = \frac{1}{2}\mathbf{U}_{ij}\theta\,x. \tag{29}$$

4.1. Frequences of free vibration and normal coordinates
To find the oscillation frequencies of the secular equation solves

$$\mid \mathbf{U} - \omega^2\mathbf{T} \mid = 0, \tag{30}$$

where, ω represents the eigenvalues, for this case, frequencies are solving the equation (30), with the obvious solutions

$$\omega_1 = \sqrt{\frac{g\,(M+m_2) + K\,l + \sqrt{g^2\,(M+m_2)^2 + K\,l\,(K\,l + 4m_1 g - 2M\,g)}}{l\,(2M - 2m_1 + m_2)}} \tag{31}$$

$$\omega_2 = \sqrt{\frac{g\,(M+m_2) + K\,l - \sqrt{g^2\,(M+m_2)^2 + K\,l\,(K\,l + 4m_1 g - 2M\,g)}}{l\,(2M - 2m_1 + m_2)}} \tag{32}$$

if the spring constant K and mass M were 0, would not determine the x variable, and the frequency ω_2 would

$$\omega_2 = \sqrt{\frac{g}{l}\frac{2m_2}{(m_2 - 2m_1)}} \tag{33}$$

The equation (33) allows you to see clearly that corresponds to an oscillator harmonic. Now if the masses m_1 and m_2 were equal to zero, the frequency ω_1 was reduced to

$$\omega_1 = \sqrt{\frac{K}{M}}, \tag{34}$$

which also corresponds to the frequency of a harmonic oscillator. Through own and knowing the eigenvalues (natural frequencies) vectors ξ normal modes of oscillation are calculated, according to this, we

$$(\mathbf{U} - \omega_2\mathbf{T})\,\mathbf{a}_k = 0 \tag{35}$$

$$\mathbf{a}^T{}_k\mathbf{T}\,\mathbf{a}_k = 1, \tag{36}$$

where,

$$\mathbf{a}_k = \begin{bmatrix} a_{1\,k} \\ a_{2\,k} \end{bmatrix} \quad with\ k = 1, 2 \tag{37}$$

$$\mathbf{A} = [\mathbf{a}_1 \quad \mathbf{a}_2]\,. \tag{38}$$

Here, $\mathbf{A}$ is the matrix of eigenvalues and equation (36) is provided orthogonality. The normal modes of oscillation are defined from the equation (39), this is

$$\boldsymbol{\xi} = \boldsymbol{A^T}\,\boldsymbol{T}\,\boldsymbol{\eta}\ , \tag{39}$$

with

$$\boldsymbol{\eta} = \begin{bmatrix} \theta \\ x \end{bmatrix} \tag{40}$$

The normal modes of oscillation are

$$\boldsymbol{\xi} = \begin{bmatrix} \frac{\sqrt{2}\,B}{2\,\sqrt{A}} \\ \\ -\frac{\sqrt{2}\,B}{2\,\sqrt{A}}, \end{bmatrix} \tag{41}$$

where A and B are defined as

$$A = 2\,M\left[g^2 + l^2{\omega_1}^4 - 2g\,l{\omega_1}^2\right] + 2m_2 g\left[g - l{\omega_1}^2\right] + 2{\omega_1}^2 m_1\,l\left[2\,g - l{\omega_1}^2\right] \tag{42}$$

$$B = 2Mx\left[g - l{\omega_2}^2\right] + m_2\left[2gx - lx{\omega_2}^2 + \theta gl\right] + 2m_1 l\left[x{\omega_2}^2 + \theta g\right]. \tag{43}$$

We have spoken so far only of vibrations along the axis; there will also be normal modes of vibration perpendicular to the axis. The complete set of normal modes is naturally more difficult to determine than the longitudinal modes, for the general motion in all directions corresponds to more degrees of freedom.

5. PID Control

Many researchers use complicated nonlinear equations or even partial differential equations as a model. But, whatever model they use, the final ontrol law should be given as a function of the acceleration of the trolley or should be derived through the dynamics of trolley motion, in Hong *et al.*[14]. In [15], proposes a Gantry crane control simulation experiment exploring the concepts of half-cycle posicast feedforward control, and Proportional, Integrative and Derivative (PID) control. Two control objectives are considered: set-point tracking and load disturbance rejection. The set-point tracking objective in this problem consists in moving the suspended load from a initial position to a specified final position. In [16], present a new procedure for designing input shapers for gantry cranes was presented. The procedure takes into account the unique properties of gantry cranes, such as, the single-mode dynamics, the known frequency range, and the standard deceleration period. The new shapers are of fixed duration, so that they give a constant deceleration period. Given the fixed shaper duration, the robustness of the shapers is maximized. The new shaping method was implemented on a gantry crane at the Savannah River Technology Center. Experimental results show that the method greatly reduces residual oscillations and closely matches the theoretically predicted performance.
To control the system described above, the spring is deleted, which exerts a force on the mass placed on the rail. The force is determined, but not as exerted by the spring, since this will be the variable being monitored. The objective is to bring the mass of the pendulum from one

point to another with a specified distance it to oscillate, so the answer is overdamped.
In [17], this article then extended to incorporate a non collocated PID controller for control of sway angle of the pendulum. Implementation results of the response of the rotary crane system with the controllers are presented in time and frequency domains. The performances of the control schemes are assessed in terms of level of sway reduction, rotational angle tracking capability and time response specifications.

To control the system described above, the spring is deleted, which exerts a force on the mass placed on the rail. The force is determined, but not as exerted by the spring, since this will be the variable being monitored. The objective is to bring the mass of the pendulum from one point to another with a specified distance it to oscillate, so the answer is overdamped.

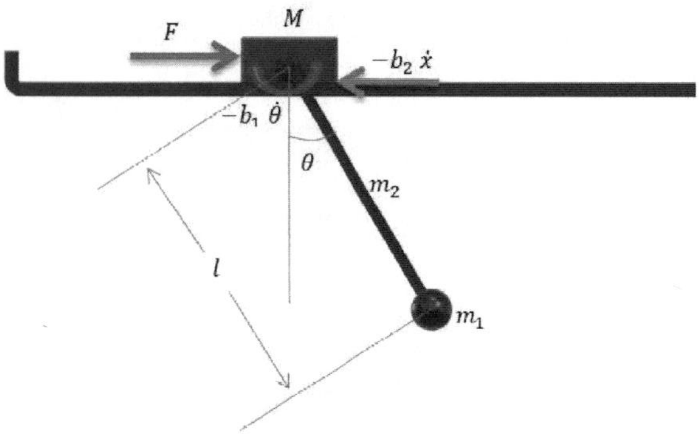

Figure 8. PID control model variables.

From the system described in Figure 8, the dynamic model is determined from the Euler-Lagrange equations for generalized force, this is

$$\frac{d}{dt}\left(\frac{\partial \mathcal{L}}{\partial \dot{q}_i}\right) - \frac{\partial \mathcal{L}}{\partial q_i} = Q_i, \qquad i = 1, 2 \tag{44}$$

where $q_1 = \theta$, $q_2 = x$, $Q_1 = Q_\theta$ and $Q_2 = Q_x$; these last two represent the generalized forces over the system, which will no longer be conservative, due to the existence of an external force F and friction is assumed in the pendulum axis $(b_1\dot{\theta})$ and carriage contact with the rail $(b_2\dot{x})$, according to Figure 8. With this variation, the new Euler-Lagrange equations, are

$$\ddot{\theta}\, D + [\ddot{x}\, l\cos(\theta) + g\, l\sin(\theta)]\; B = -b_1\dot{\theta} \tag{45}$$

$$\ddot{x}\, C + [\ddot{x}\, l\cos(\theta) + g\, l\sin(\theta)]\; B = F - b_2\,\dot{x} \tag{46}$$

where

$$D = \left(\frac{2}{5} m_1\, r^2\right), \tag{47}$$

and B, C are the same terms in the equations (17) and (18), respectively. To control the system, you must first find the state space,

$$\mathbf{Z} = \left\{ \begin{array}{c} \theta \\ \dot{\theta} \\ x \\ \dot{x} \end{array} \right\}, \tag{48}$$

such that the derivative of $\mathbf{Z}$ is

$$\dot{\mathbf{Z}} = \left\{ \begin{array}{c} Z_2 \\ \dfrac{l\,B\cos(Z_1)\,\epsilon_1 + C\,\epsilon_2}{\epsilon_3} \\ Z_4 \\ \dfrac{\left(l^2\,B + D\right)\,\epsilon_1 + \left(l\,B\cos(Z_1)\right)\,\epsilon_2}{-\epsilon_3} \end{array} \right\}, \tag{49}$$

where

$$\begin{aligned} \epsilon_1 &= F - b_2\dot{x} + \dot{\theta}^2 l\,B\sin(Z_1), \\ \epsilon_2 &= b_1\dot{\theta} + g\,l\,B\sin(Z_1), \\ \epsilon_3 &= l^2\,B^2\cos^2(Z_1) - l^2\,B\,C - D\,C. \end{aligned}$$

The equations in space of states (49) were implemented in Simulink to obtain the natural response of the system to a step input in F and subsequently draw the desired trajectory. In the scheme of Simulink, the value for $b_1 = 0.1$, and $b_2 = 0.5$, was assumed; because the system is dimensionally small, the input for the force F was 0.1, for a time of 2 seconds. The natural response to these values are shown in Figure 9 and Figure 10 for θ, x respectively.

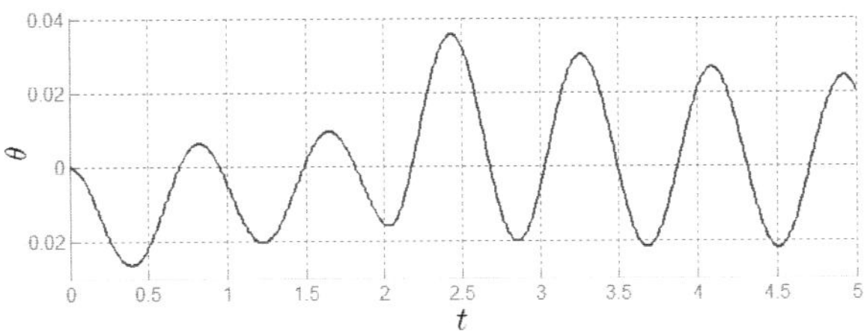

Figure 9. Output to an input $F = 0.1$, for 2 seconds.

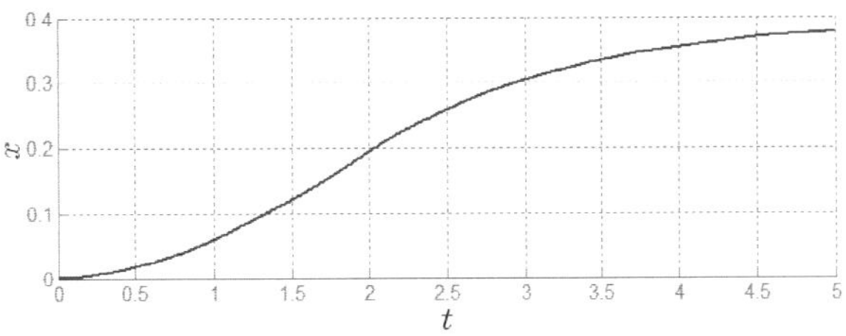

Figure 10. Output to an input $F = 0.1$, for 2 seconds.

A suitable trajectory for the variable x is that in which the speed starts and ends at zero, the trigonometric function tanh, so that the function of the desired trajectory is,

$$x^* = \frac{d}{2}\tanh\left[\alpha\left(t - \frac{T}{2}\right)\right] + \frac{d}{2}, \tag{50}$$

where d is the distance to travel, t represents time, T the time and α a factor to vary the slope of the curve. The function type proposed is more than twice differentiable, this ensures a continuous acceleration along the way. Having defined the trajectory to x, these values are entered into equation (45), and trajectories for θ y $\dot{\theta}$ are found. To the controller of the variable θ, the tuned values are

Table 3. Controller θ

Conditions		Value
Proportional	=	3
Integral	=	0.25
Derivative	=	0.5

IOP Conf. Series: Journal of Physics: Conf. Series **850** (2017) 012004 doi:10.1088/1742-6596/850/1/012004

and for the variable x, we are

Table 4. Controller x

Conditions		Value
Proportional	=	0.01
Integral	=	1.7
Derivative	=	2

The Figure 11 shows the block diagram for this system,

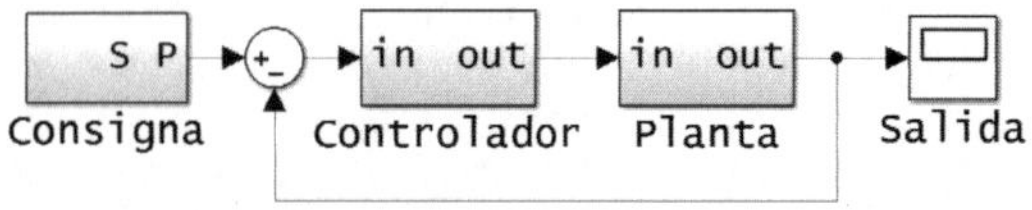

Figure 11. Block diagram of the system with the incorporated controller.

in which the controller is already implemented, which in turn has two PID, because they are two variables which require control. This diagram represents the general model of the system with the implemented controller, in which the *Planta*(Plant) block represents the mechanical system, or the grantry crane. From this are taken readings of angular position of the pendulum and linear position of the car, as well as their respective speeds and accelerations. These output parameters are subtracted from the *Consigna*(Set point) and the error is determined (which is simply the subtraction of the desired output with the actual output), which becomes the PID controller input, sending a power signal to the engine that moves the system (Plant).

The output for $\theta, \dot{\theta}$ and $x, \dot{x}$ is registered in Figure 12 and Figure 13, respectively

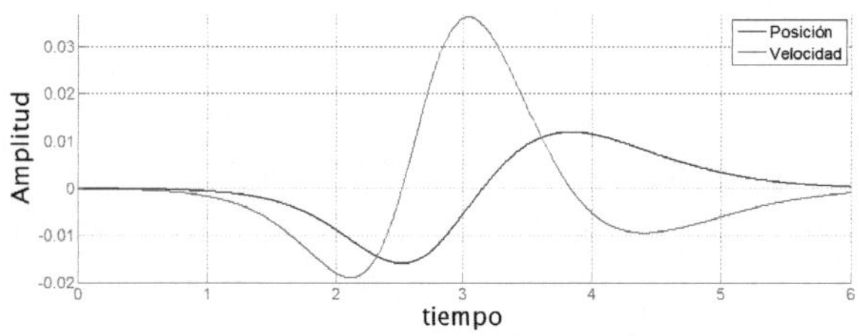

Figure 12. Output for θ and $\dot{\theta}$.

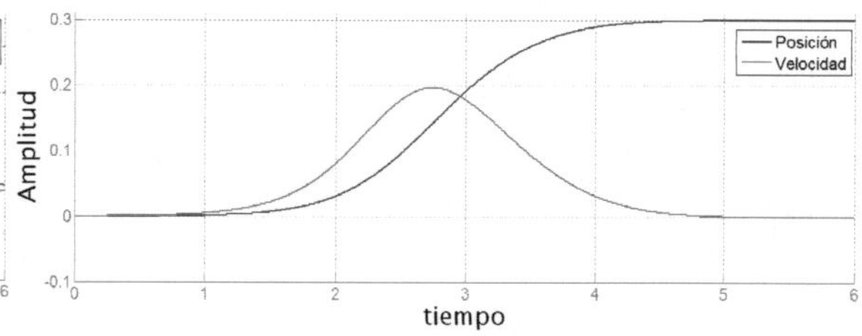

Figure 13. Output for x and $\dot{x}$.

As seen in the previous two graphs, variables were controlled according to the desired trajectory. In addition to the equations included in Simulink, the CAD model was implemented in SimMechanics, with the purpose of verify the controller system, which is made use of SimMechanics-link, which exports SolidWorks CAD model to MATLAB. As for the position and velocity of the pendulum shown in Figure 12, it can be seen that the controller tries to keep the pendulum at its equilibrium point, It can be seen in the red line that the position has few oscillations, so that remains in cuasi-equilibrium. The Figure 13 shows the position and speed of the cart under the action of the controller, it is observed that the cart arrives at the desired position by the trajectory defined in a non-abrupt way and without going back. As for the speed, we see that this increases when the cart is starting and decreases when it is reaching its stationary zone, so that the car remains with a null position error.

6. Conclusions

The Euler-Lagrange formalism allows dynamic modeling of crane-type system in a simple way, thanks to the formalism is based on energy principles. This paper has been presented step by step the dynamic modeling of the proposed system, in addition to the respective simulation of the solution of the equations and later validation of the results using a CAD model simulation designed in SolidWorks in the extension of SimMechanics Matlab. This document differs from others through the use of SimMechanics extension for simulation CAD model, this tool is simple to use and reliable, this is anadvantage for simulations of 3D model since the creation of platforms is not necessary for their simulation. The results were represented graphically and are successful, since two different ways came to the same results with a visually low error between the two procedures. Additionally, in this particular article analysis it is made as to the desired trajectories, such phase space analysis is topologically correct, this being a good contribution to future dynamic studies, kinematic and physical of the system. The development of this work requires a kind of dynamic modelling to study the properties needed for monitoring tasks using techniques associated with control. Consists in know thoroughly the system under study, the context in which it is located, meet its internal dynamics, variables, its operation, its components and the relationships between them, the maximum error to tolerate, alternate models are known, the point(s) of desired operation, the operating range, etc. In addition, it must satisfy the objectives and expectations of the dynamic to develop, among other things of the model. The PID is a type of controller in state space that responds well to perturbations. In addition, this driver guarantees us a null position error. The controller responds correctly, since it takes the carriage to the desired position in a good way following the path defined in (50). Another step is the characterization of the variables and equations of motion. In this phase the dynamic model of the proposed systems was studied, for which the Euler-Lagrange formalism was used because it fits perfectly for the study of nonlinear dynamical systems. After having the motion equations of the systems, it proceeded to solve the differential equations obtained, which are partial, of the second order, and coupled. The phase control, PID control was used, with which we obtained a synthesis of the drivers from the equations of motion but in the space of states, obtained in the characterization phase, it was possible to represent dynamic systems as a whole of coupled differential equations of first order. In this representation, known as representation in the state space, it shows the variation of a vector z, known as state vector and contains the most important system information, as a function of the same z vector and an input vector u. From this formulation, it was possible to synthesize laws nonlinear control.

6.1. Acknowledgments

Authors want to thank the financial support from Mathematical School from Universidad Industrial de Santander (UIS) and Universidad Santo Tomás.

References

[1] Abdel-Rahman EM, Nayfeh AH, Masoud ZN. *Dynamics and control of cranes: A review.* J Vib Control 2001; **9**: 863-908.

[2] Jones JF, Patterson BJ. Oscillation damped movements of suspended payloads. In: *Proceedings of the IEEE international conference on robotics and automation*; 1988.

[3] Henry EJ, Masoud ZN, Nayfeh AH, Mook DT. *Cargo pendulation reduction on ship-mounted cranes via boom-luff angle actuation.* J Vib Control 1999; **7**: 1253-64.

[4] Ma BJ, Fang YC, Liu XE, Wang PC. *Modeling and simulation platform design for 3d overhead crane.* J Syst Simul 2009; **21**: 3798-803.

[5] Cartmell MP. *On the need for control of nonlinear oscillations in machine systems.* Mecanica 2003; **38**: 185-212.

[6] Cartmell MP, Morrish L, Alberts TE, Taylor AJ. *Controlling the nonlinear dynamics of gantry cranes.* Mach Vib 1996; **5**: 197-210.

[7] G.A. Manson, "*Time-Optimal Control of and Overhead Crane Model,*" Optimal Control Applications and Methods, **3**(2), 1992, pp. 115-120.

[8] J. Auernig and H. Troger, "*Time Optimal Control of Overhead Cranes with Hoisting of the Load,*" Automatica, **23**(4), 1987, pp. 437-447.

[9] H.M. Omar, Control of Gantry and Tower Cranes, M.S. Thesis, Virginia Tech., 2003, Blacksburg, VA.

[10] H.H. Lee and S.K. Cho, "*A New fuzzy logic anti-swing control for industrial three-dimensional overhead crane,*" Proceedings of the 2001 IEEE International Conference on Robotic and Automation, 2001, Seoul, pp. 2958-2961.

[11] Meirovitch L. *Methods of analytical dynamics.* New York: McGraw-Hill, Inc.; 1988.

[12] Goldstein. *Classical Mechanics.* Editorial Addison Wesley (1980).

[13] O. Sawodny, H. Aschemann and S. Lahres, "*An automated gantry crane as a large workspace robot,*" Control Engineering Practice, Vol. **10**, No. 12, pp. 1323-1338, 2002.

[14] Yong-Seok Kim, Keum-Shik Hong, and Seung-Ki Sul. *Anti-Sway Control of Container Cranes: Inclinometer, Observer, and State Feedback* International Journal of Control, Automation, and Systems, 2004; **4**: 435-439.

[15] P. B. de Moura Oliveira, J. Boaventura Cunha. *Gantry Crane Control: a Simulation Case Study.* IEEE, 2013: 978-1-4799-2741-8/13

[16] Neil Singer, William Singhose, Eric Kriikku. *An Input Shaping Controller Enabling Cranes to Move Without Sway.* American Nuclear Society 7th Topical Meeting on Robotics and Remote Systems

[17] M.A. Ahmad, M.S. Ramli, M.A. Zawawi, R.M.T. Raja Ismail. *Hybrid Collocated PD with Non-collocated PID for Sway Control of a Lab-scaled Rotary Crane.* IEEE, 2010: 978-1-4244-5046-6/10

Dynamic Modeling and Simulation of an Underactuated System

Juan Libardo Duarte Madrid[1], P A Ospina Henao[2] and E González Querubín[3]

[1] Facultad de Ingeniería Mecatrónica, Universidad Santo Tomás, Bucaramanga, Colombia.
[2] Departamento de Ciencias Básicas, Universidad Santo Tomás, Bucaramanga, Colombia.
[3] Facultad de Ingeniería Mecatrónica, Universidad Santo Tomás, Bucaramanga, Colombia.

E-mail: juan.duarte03@ustabuca.edu.co, paolo.ospina@ustabuca.edu.co, edwin.gonzalez@ustabuca.edu.co

Abstract. In this paper, is used the Lagrangian classical mechanics for modeling the dynamics of an underactuated system, specifically a rotary inverted pendulum that will have two equations of motion. A basic design of the system is proposed in SOLIDWORKS 3D CAD software, which based on the material and dimensions of the model provides some physical variables necessary for modeling. In order to verify the results obtained, a comparison the CAD model simulated in the environment SimMechanics of MATLAB software with the mathematical model who was consisting of Euler-Lagrange's equations implemented in Simulink MATLAB, solved with the ODE23tb method, included in the MATLAB libraries for the solution of systems of equations of the type and order obtained. This article also has a topological analysis of pendulum trajectories through a phase space diagram, which allows the identification of stable and unstable regions of the system.

1. Introduction

Nowadays, the underactuated mechanical systems [1] are generating interest among researchers of modern control theories. This interest is that these systems have similar problems to those found in industrial applications, such as disturbances, instabilities and nonlinear behavior in some conditions of operation. The rotary inverted pendulum [2] is a clear example of an underactuated mechanical system; this is a mechanism of two degrees of freedom (DOF) and two rotational joints. It consists of three main elements: a motor, a rotational arm and a pendulum [3]. The motor shaft is connected to one end of the rotational arm making this fully rotate on a horizontal plane; the other end of the arm has connected the pendulum that freely rotates 360 degrees in a vertical plane. Despite being a purely academic level, this system is helpful to study, apply and analyze different modeling strategies.

Some industrial applications [4] that present the disadvantages mentioned above, are in fields such as robotics, robots balance, biped robots [5], robotic arms; telecommunications, satellite positioning; transportation, Segway [6], [7], stability of ships and submarines, iBot [8], Self-balancing unicycle; and field monitoring, drones.

For the accomplishment of this work, initially there was realized a bibliographic review of similar projects [9], [10], [11], in order to determine the dimensions and suitable material for the later design of the prototype in SOLIDWORKS 3D Computer Aided Engineering (CAD) software. Then, was modeled the dynamics of the system by Lagrangian mechanics, obtaining two equations of motion; one corresponds to the arm and other one to the pendulum. To validate the equations obtained, multiple simulations were made in the MATLAB software in order to observe their behavior. Thus, it was possible to verify in a graphic mode, the waveform of the angular position and velocity of the arm and the pendulum.

A fundamental part in the stage of simulations was to import the CAD model made in SOLIDWORKS to environment of SimMechanics [12] in MATLAB. Immediately afterwards, the equations obtained were solved by the ODE23tb (Ordinary Differential Equations) method belonging to MATLAB. The use of block diagram in the environment Simulink of MATLAB allowed a representation of the equations of motion. All this had the purpose of elaborating a comparison of the response of both the CAD model and the mathematical model obtained by the Lagrange formalism.

Finally, through the effective potential of the system the phase space diagram [13] was plotted, in this one can see the trajectories of the pendulum and to study the critical points that correspond to the maximum and minimum of that potential.

In Section II is defined in a broader way the rotational inverted pendulum, in addition a table is shown that contains some variables that will be used throughout the article along with a basic design of the system. In section III It provides a brief definition of an underactuated systems. Next, in section IV presents the advantage of making use of Euler-Lagrange formalism. Continuing, in Section V it shows the dynamic modeling of the proposed rotary inverted pendulum. In section VI activities are defined for performing proposed simulations as a method of validating results. In section VII the results obtained are shown with a respective analysis of them. Finally, section VIII contains the relevant conclusions of the article.

2. Prototype Proposed

The pendulum is a physical object consisting of a mass joined by means of a bar to a pivot. The system is free to move, realizing a balancing about its point of stable equilibrium; namely, in its hanging position, in which it will be as long as there are no forces that influence its nature. In 1991 [14], there was created the system that shows itself in the Figure 1, this device is known as Furuta's pendulum, referring to your creator Katsuhisa Furuta. This system consists of two links only, the first one of them is known as arm, which is an actuated element. Since this one turns freely in a horizontal plane with the help of an electrical engine of direct current, the second one knows himself as pendulum; this one is an underactuated element that turns freely in a vertical plane for action of the movement transmitted by the arm. Therefore, it is a question of an underactuated system with two GDL and one control input [15].

The rotary inverted pendulum or Furuta's pendulum is of great help in the academy for the analysis and study of strategies of classic, modern, not linear and advanced control; also, it is used with didactic intentions for the education of the theory of control, classic mechanics and system identification.

This section presents the basic design created in SOLIDWORKS and some variables needed for modeling.

Table 1. Variable System

Physical characteristic	Symbology	Units
Acceleration of gravity	g	m/s^2
Angular position of the arm	θ_0	rad
Angular position of the pendulum	θ_1	rad
Arm length	L_0	m
Generalized coordinates	q	rad

Generalized velocities	$\dot{q}$	rad/s
Kinetic Energy	K	J
Lagrangian	$\mathcal{L}$	J
Location of the center of mass of the pendulum	l_1	m
Moment of inertia of the arm	I_0	$kg.m^2$
Moment of inertia of the pendulum	I_1	$kg.m^2$
Pendulum length	L_1	m
Pendulum mass	m_1	kg
Potential Energy	U	J

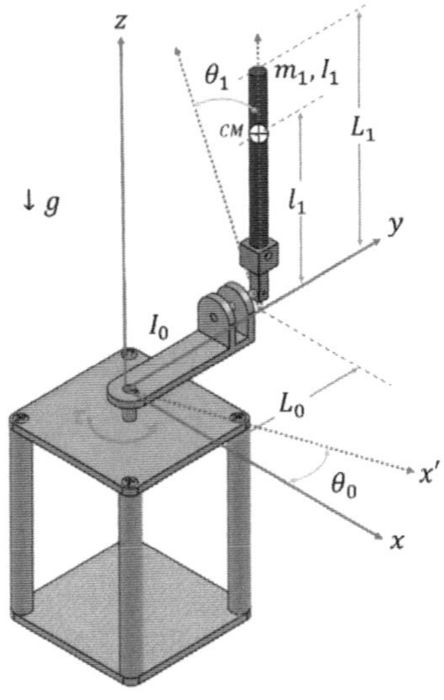

Figure 1. Rotary inverted pendulum

3. Underactuated Systems

In the study of mechanisms are acquired two concepts very important such as the direct and indirect action. The first consists of movement of elements by action of an actuator, while the second consists of the action of motion transmitted by another interconnected element. Such movements are known as DOF, so that mechanical systems or mechanisms can be classified depending on the number of DOF and the number of actuators. The fully actuated mechanical systems are those having the same number of DOF and actuators. Underactuated mechanical systems are those with fewer actuators than DOF [16]. It is important to highlight the advantages of underactuated systems, since if they do not have advantages over fully actuated mechanical systems, it will not make sense its development. The main advantages present in underactuated systems are energy saving and control efforts. However, these systems are intended to perform the same functions of fully actuated systems without their disadvantages.

4. Euler-Lagrange

The dynamic equations of any mechanical system can be obtained from the known Newtonian classical mechanics, the drawback of this formalism is the use of the variables in vector form, complicating considerably the analysis when increasing the joints or there are rotations present in the system. In these cases, it is favorable to employ the Lagrange equations, which have formalism of scale, facilitating the analysis for any mechanical system.

In order to use Lagrange equations, it is necessary to follow four steps:

- Calculation of kinetic energy.
- Calculation of the potential energy.
- Calculation of the Lagrangian.
- Solve the equations.

Where the kinetic energy can be both rotationally and translational, this form of energy may be a function of both the position and the speed $K\big(q(t),\dot{q}(t)\big)$.

The potential energy is due to conservative forces as the forces exerted by springs and gravity, this energy is in terms of the position $U(q(t))$.

Is defined the Lagrangian as

$$\mathcal{L} = K - U \tag{1}$$

Therefore, the Lagrangian in general terms is defined of the following way

$$\mathcal{L}\big(q(t),\dot{q}(t)\big) = K\big(q(t),\dot{q}(t)\big) - U\big(q(t)\big) \tag{2}$$

Finally, are defined the Euler-Lagrange equations for a system of n DOF as follows

$$\frac{d}{dt}\left(\frac{\partial\mathcal{L}(q,\dot{q})}{\partial\dot{q}_i}\right) - \frac{\partial\mathcal{L}(q,\dot{q})}{\partial q_i} = \tau_i \tag{3}$$

Where $i = 1,\dots,n$, τ_i are the forces or pairs exerted externally in each joint, besides nonconservative forces such as friction, resistance to movement of an object within a fluid and generally those that depend on time or speed. It will be obtained an equal number of dynamic equations and DOF.

5. Modeling of Rotary Inverted Pendulum

Was performed the dynamic modeling of the model shown in the Figure 1. It is necessary to make an energy analysis. Therefore, initially the kinetic energy of each link is analyzed, so it can be identified which kinetic energies (rotational and translational) were present in each link.

5.1. Kinetic Energy

The kinetic energy of the system consist of a translational and rotational component for the pendulum and rotational component for the arm

$$K = \frac{1}{2}mv^2 + \frac{1}{2}I\omega^2 \tag{4}$$

Where m is the mass of the body, v the linear velocity, I the moment of inertia, ω the angular velocity and K the kinetic energy. In this case, there are two bodies, the arm and the pendulum. The kinetic energy of the arm is

$$K_0 = \frac{1}{2}I_0\dot{\theta_0}^2 \tag{5}$$

The kinetic energy of the pendulum is

$$K_1 = \frac{1}{2} m_1 v_1{}^2 + \frac{1}{2} I_1 \dot{\theta}_1{}^2 \tag{6}$$

The total energy of the system is

$$K_T = \frac{1}{2} I_0 \dot{\theta}_0{}^2 + \frac{1}{2} m_1 v_1{}^2 + \frac{1}{2} I_1 \dot{\theta}_1{}^2 \tag{7}$$

5.2. Potential Energy
This system only store gravitational potential energy in the pendulum

$$U = mgh \tag{8}$$

The arm has in its nature a rotational movement in a horizontal plane, therefore it does not have height change in its center of mass, providing equal 0 component in equation (8) resulting in a potential energy zero .
The potential energy of the pendulum is

$$U_1 = m_1 g l_1 (\cos\theta_1 - 1) \tag{9}$$

Where g represents the value of gravity. The total potential energy of the system is (9)

$$U_T = m_1 g l_1 (\cos\theta_1 - 1) \tag{10}$$

5.3. Position of the pendulum
Because the pendulum is a rigid body, the required position is the its center of mass

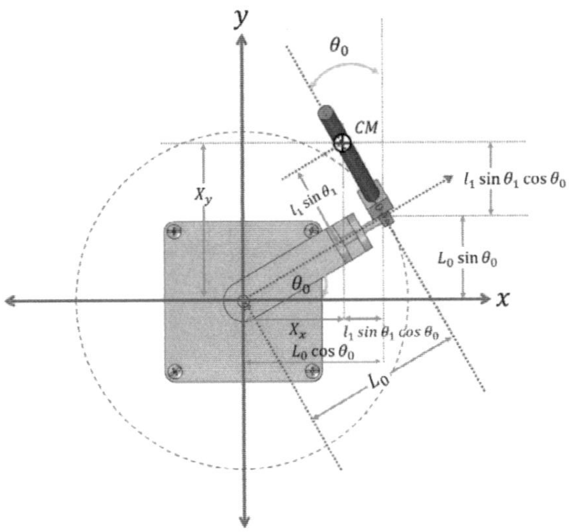

Figure 2. Projection arm and pendulum in the xy plane

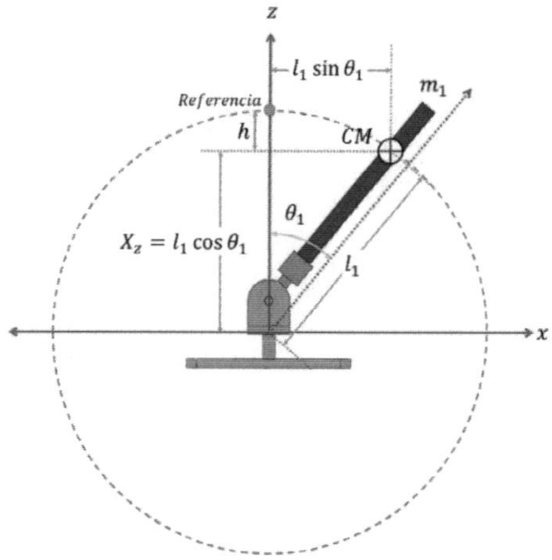

Figure 3. Projection pendulum in the xz plane

Then, it shows components the position of the center of mass

$$X_x = L_0 \cos\theta_0 - l_1 \sin\theta_1 \sin\theta_0 \tag{11}$$

$$X_y = L_0 \sin\theta_0 + l_1 \sin\theta_1 \cos\theta_0 \tag{12}$$

$$X_z = l_1 \cos\theta_1 \tag{13}$$

The position of the center of mass

$$X_{CM} = [X_x \quad X_y \quad X_z]^T \tag{14}$$

5.4. Linear Speed of the center of mass

The speed is defined as the derivative of the position with respect to time

$$\frac{d}{dt}(X_{CM}) = V_{CM} \tag{15}$$

$$\dot{X}_{CM} = V_{CM} = \left[\dot{X}_x \quad \dot{X}_y \quad \dot{X}_z\right]^T \tag{16}$$

The velocity components are obtained by differentiating each component position (11), (12) and (13) respectively

$$\dot{X}_x = -\dot{\theta}_0 L_0 \sin\theta_0 - l_1(\dot{\theta}_0 \sin\theta_1 \cos\theta_0 + \dot{\theta}_1 \cos\theta_1 \sin\theta_0) \tag{17}$$

$$\dot{X}_y = \dot{\theta}_0 L_0 \cos\theta_0 + l_1(\dot{\theta}_1 \cos\theta_1 \cos\theta_0 - \dot{\theta}_0 \sin\theta_1 \sin\theta_0) \tag{18}$$

$$\dot{X}_z = -\dot{\theta}_1 l_1 \sin\theta_1 \tag{19}$$

The linear velocity of the center of mass can be expressed as follows vector form

$$V_{CM}{}^2 = \begin{bmatrix} \dot{X}_x & \dot{X}_y & \dot{X}_z \end{bmatrix} \begin{bmatrix} \dot{X}_x \\ \dot{X}_y \\ \dot{X}_z \end{bmatrix} = \dot{X}_x{}^2 + \dot{X}_y{}^2 + \dot{X}_z{}^2 \tag{20}$$

They are calculated separately each of the components to the square of the velocity of the center of mass

$$\dot{X}_x{}^2 = \dot{\theta}_0{}^2 L_0{}^2 \sin^2\theta_0 - 2(\dot{\theta}_0 L_0 \sin\theta_0)\left(l_1(\dot{\theta}_0 \sin\theta_1 \cos\theta_0 + \dot{\theta}_1 \cos\theta_1 \sin\theta_0)\right) + l_1{}^2(\dot{\theta}_0 \sin\theta_1 \cos\theta_0 + \dot{\theta}_1 \cos\theta_1 \sin\theta_0)^2 \tag{21}$$

$$\dot{X}_y{}^2 = \dot{\theta}_0{}^2 L_0{}^2 \cos^2\theta_0 + 2(\dot{\theta}_0 L_0 \cos\theta_0)\left(l_1(\dot{\theta}_1 \cos\theta_1 \cos\theta_0 - \dot{\theta}_0 \sin\theta_1 \sin\theta_0)\right) + l_1{}^2(\dot{\theta}_1 \cos\theta_1 \cos\theta_0 - \dot{\theta}_0 \sin\theta_1 \sin\theta_0)^2 \tag{22}$$

$$\dot{X}_z{}^2 = \dot{\theta}_1{}^2 l_1{}^2 \sin^2\theta_1 \tag{23}$$

Finally, the following expression is obtained for the linear velocity

$$V_{CM}{}^2 = L_0{}^2\dot{\theta}_0{}^2 + l_1{}^2(\dot{\theta}_1{}^2 + \dot{\theta}_0{}^2 \sin^2\theta_1) + 2L_0 l_1 m_1 \dot{\theta}_0 \dot{\theta}_1 \cos\theta_1 \tag{24}$$

5.5. Energy System

The total energy of the system is

$$K_T = K_0 + K_1 \tag{25}$$

Substituting (5) and (6) in (25) is obtained

$$K_T = \frac{1}{2} I_0 \dot{\theta}_0{}^2 + \frac{1}{2} I_1 \dot{\theta}_1{}^2 + \frac{1}{2} m_1 V_{CM}{}^2 \tag{26}$$

Substituting (24) in (26) is obtained the total kinetic energy

$$K_T = \frac{1}{2} I_0 \dot{\theta}_0{}^2 + \frac{1}{2} I_1 \dot{\theta}_1{}^2 + \frac{1}{2} m_1 \{L_0{}^2\dot{\theta}_0{}^2 + l_1{}^2(\dot{\theta}_1{}^2 + \dot{\theta}_0{}^2 \sin^2\theta_1) + 2L_0 l_1 m_1 \dot{\theta}_0 \dot{\theta}_1 \cos\theta_1\} \tag{27}$$

Is shown in (10) the potential energy of the system

5.6. Euler-Lagrange Equations

The Lagrangian of the system is

$$\mathcal{L} = K_T - U_T \tag{28}$$

Substituting (10) and (27) in (28) the Lagrangian of the system is obtained

$$\mathcal{L} = \frac{1}{2} I_0 \dot{\theta}_0{}^2 + \frac{1}{2} I_1 \dot{\theta}_1{}^2 + \frac{1}{2}\left(L_0{}^2 m_1 \dot{\theta}_0{}^2\right) + \frac{1}{2}\left(l_1{}^2 m_1 \dot{\theta}_1{}^2\right) + \frac{1}{2}\left(l_1{}^2 m_1 \dot{\theta}_0{}^2 \sin^2\theta_1\right) + L_0 l_1 m_1 \dot{\theta}_0 \dot{\theta}_1 \cos\theta_1 + m_1 g l_1 (1 - \cos\theta_1) \tag{29}$$

As there are two DOF, the Euler-Lagrange equations have the following form

$$\frac{d}{dt}\left(\frac{\partial \mathcal{L}}{\partial \dot{\theta}_0}\right) - \frac{\partial \mathcal{L}}{\partial \theta_0} = \tau \tag{30}$$

$$\frac{d}{dt}\left(\frac{\partial \mathcal{L}}{\partial \dot{\theta}_1}\right) - \frac{\partial \mathcal{L}}{\partial \theta_1} = 0 \tag{31}$$

Where τ is the torque of motor, solving (30) and (31) we obtain the equations of motion which are given by

$$I_0\ddot{\theta}_0 + L_0{}^2 m_1\ddot{\theta}_0 + \left\{l_1{}^2 m_1\left(\ddot{\theta}_0 \sin^2\theta_1 + 2\dot{\theta}_0\dot{\theta}_1 \sin\theta_1 \cos\theta_1\right)\right\} + \left\{L_0 l_1 m_1\left(\ddot{\theta}_1 \cos\theta_1 - \dot{\theta}_1{}^2 \sin\theta_1\right)\right\} = \tau \tag{32}$$

$$I_1\ddot{\theta}_1 + l_1{}^2 m_1\ddot{\theta}_1 + L_0 l_1 m_1\ddot{\theta}_0 \cos\theta_1 - l_1{}^2 m_1\dot{\theta}_0{}^2 \sin\theta_1\cos\theta_1 - m_1 g l_1 \sin\theta_1 = 0 \tag{33}$$

Where (32) is the equation of motion of the arm and (33) of the pendulum.

6. Simulation

Although the Euler-Lagrange formalism ensures a high degree of approximation of the mathematical models, is essential do comparisons to validate these results. Are followed the next steps for verification of modeling:

1) Represent the system equations in the state space.
2) Define an experiment with initial conditions, natural interactions and external forces.
3) Import the CAD model of SOLIDWORKS in SimMechanics-MATLAB.
4) Add the necessary blocks to obtain the desired graphic model and applying external forces.
5) Simulating experiment.
6) Export the results of SimMechanics to Workspace MATLAB.
7) Implement a block diagram Simulink-MATLAB to solve the equations.
8) Export the solutions to the equations to Workspace MATLAB.
9) Graphing and overlay solutions.

As can be seen in the steps above, the simulation of the model was divided in two stages: first, simulate the CAD model initially designed and the second implement the equations obtained.

7. Results

Then, it presents each of the steps mentioned in the previous section.

1) Starting from [17]

$$M(q)\ddot{q} + C(q,\dot{q})\dot{q} + G(q) = \tau \tag{34}$$

Therefore, the equation (34) is the dynamic equation for mechanical systems of n DOF. Where, M is the matrix of inertia, C is the centrifugal and Coriolis matrix, G the vector of gravity and τ external forces.

Taking the equations of motion (32) and (33) and replacing in (34) is obtained a matrix representation

$$\begin{bmatrix} I_o + m_1 L_0{}^2 + l_1{}^2 m_1 \sin^2\theta_1 & L_0 l_1 m_1 \cos\theta_1 \\ L_0 l_1 m_1 \cos\theta_1 & I_1 + m_1 l_1{}^2 \end{bmatrix}\begin{bmatrix}\ddot{\theta}_0 \\ \ddot{\theta}_1\end{bmatrix} + \begin{bmatrix} 2l_1{}^2 m_1 \sin\theta_1 \cos\theta_1\,\dot{\theta}_1 & -L_0 l_1 m_1 \sin\theta_1\,\dot{\theta}_1 \\ -l_1{}^2 m_1 \sin\theta_1\cos\theta_1\,\dot{\theta}_0 & 0 \end{bmatrix}\begin{bmatrix}\dot{\theta}_0 \\ \dot{\theta}_1\end{bmatrix} + \begin{bmatrix} 0 \\ -g l_1 m_1 \sin\theta_1 \end{bmatrix} = \begin{bmatrix}\tau \\ 0\end{bmatrix} \tag{35}$$

The matrix $M(q)$ is important for the dynamic modeling and for the design of controllers. This matrix has a great relationship with the kinetic energy, also the inertia matrix is a symmetric, positive and square matrix of $n\times n$, whose elements depend only on the generalized coordinates.

Centrifugal and Coriolis matrix $C(q,\dot{q})$ it is important in the study of stability in control systems, mechanical systems, among others. This matrix is square of $n\times n$ and has dependence in its elements of the generalized coordinates and velocities.

The gravity vector $G(q)$ it is present in mechanical systems without counterweights or springs, in turn is in systems with displacement off the horizontal plane. This vector is of $n\times 1$ and has only reliance on joint positions.

2) It defined that the system would have the initial conditions shown in Figure 5, in addition to being subject to effects of gravity and to a torque step 0.2 seconds in the end of the arm that connects to the motor shaft. Finally, a simulation interval 5 seconds was established.

3) Was imported the CAD model in SimMechanics with the following code line: *mech_import('CADModel_Pendulum.xml')*. The initial position of the system is Figure 5 a).

4) It was necessary add a few block the diagram SimMechanics, since the CAD model is only under the effect of gravity and not have some kind of movement, it blocks provide the step of torque arm included to start rotating, besides adding blocks to the sensing of angular displacement and velocities in an interval of 5 second of test.

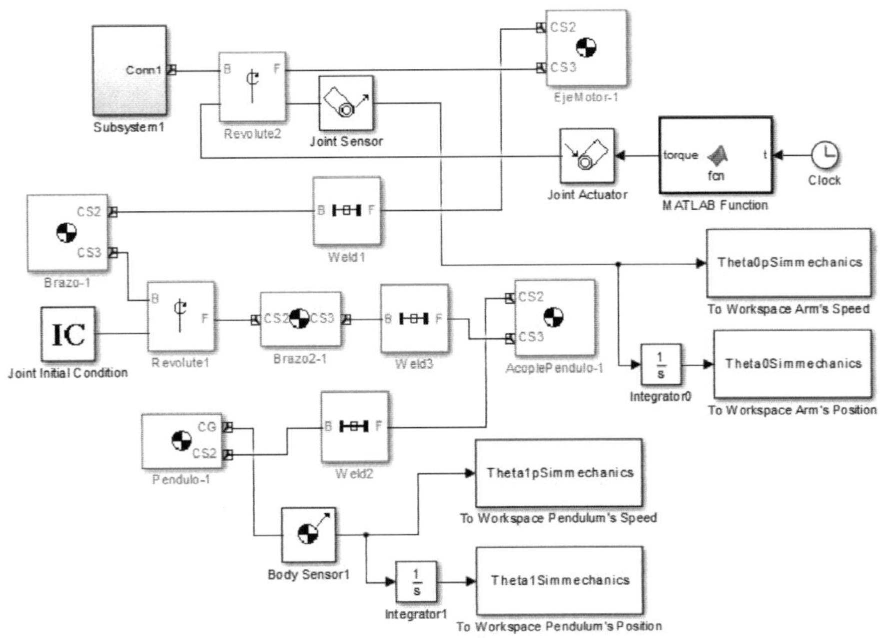

Figure 4. Final block diagram in SimMechanics

The Figure 4 shows the block diagram implemented for the simulation of the CAD model. The block named Subsystem1 contains the majority of the blocks generated of the CAD model by SimMechanics, such blocks are not necessary for the simulation; others blocks generated by SimMechanics are Revolute2, EjeMotor-1, Weld1, Brazo-1, Revolute1, Brazo2-1, Weld3, AcoplePendulo-1, Weld2 and Pendulo-1. The most important elements for the simulation are the blocks Revolute; Revolute2 is the joint of the arm, theoretically it is there where there would be connected the electrical engine, which in this case is simulated by a Joint Actuator configured as a force generator. The block MATLAB Function contains a small code that generates a torque of 0.5 Nm in an interval of 0.2 seconds; to simulate the motor encoder the Joint Sensor block is connected, of which there is obtained the speed of arm it $\dot{\theta}_0$ (Figure 11), and with an of integration block (Integrator0) the position of same one it θ_0 (Figure 9). For the case of the Revolute1 there is a bit different the way of obtaining the variables of interest, this difference owes to the CAD design in SOLIDWORKS, since both the arm and the pendulum have design and couplings different. In Revolute1 there connects a block called Joint Initial Condition, in which it is possible to change the initial position of the pendulum, in this link we will not have any engine, since as mentioned in the section II it is a question of the underactuated element. On the other hand, it is necessary to simulate an encoder to know the position of the pendulum in any instant of time, for this implemented the block Body Sensor1, which is used to measure the position of the center of mass of the pendulum. So that, there is obtained initially the angular speed of the pendulum it $\dot{\theta}_1$ (Figure 10) and of equal way that with the arm, by means of an integrator (Integrator1) obtains the angular position it θ_1 (Figure 8).

5) Some moments of the position of the system were taken in the time interval defined.

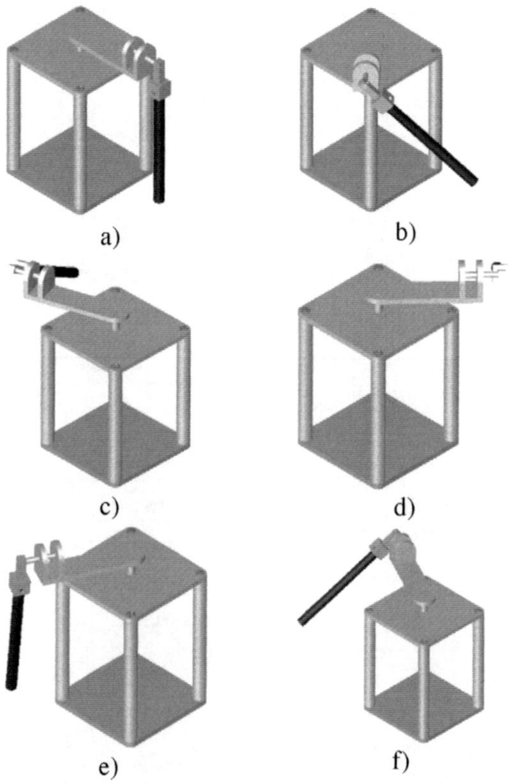

Figure 5. Experiment in SimMechanics
a) t = 0 s, b) t = 0.212227 s, c) t = 0.798791 s, d) t = 1.22854 s, e) t = 3.33321 s, f) t = 5 s

6) To export system solutions to Workspace, is need to add the blocks with the name To Workspace located in Simulink library, which will create a cell for each solution, with the respective data and time in which such data is obtained.

7) The implementation of the block diagram was performed to numerically solve the equations of the system, thus obtaining values $\theta_0, \theta_1, \dot{\theta}_0, \dot{\theta}_1$ along the defined time interval. The diagram implemented is the follows:

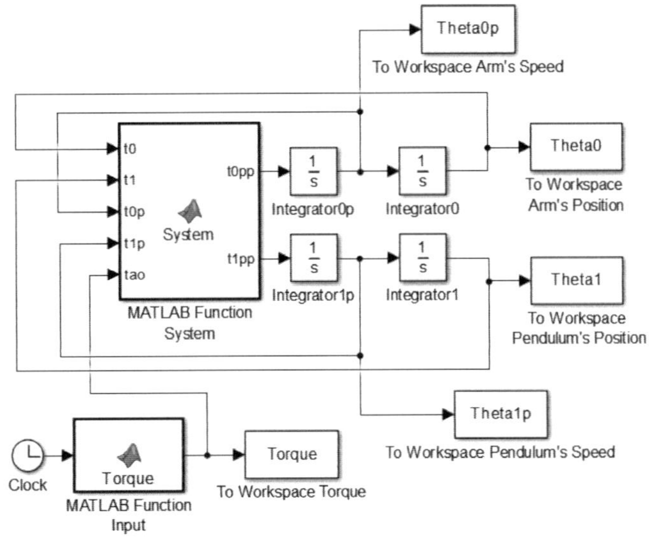

Figure 6. Block diagram of Euler-Lagrange's equations

Wherein, the block MATLAB Function Input, has the input applied to the system, which in this case is a torque step of 0.2 seconds with amplitude of 0.5 Nm, this being the maximum torque generated by the motor. The block called MATLAB Function System, contains the mathematical model in its matrix representation, with five inputs $(\theta_0, \theta_1, \dot{\theta}_0, \dot{\theta}_1, \tau)$ and two outputs $(\ddot{\theta}_0, \ddot{\theta}_1)$. The integrators are used to obtain speed and position respectively. Internally, the block uses the ODE23tb method for solving equations.

Numerical values necessary to solve the Euler-Lagrange's equations of motion are as follows

Table 2. Numerical Variable System

Physical characteristic	Values
Acceleration of gravity	$9.81\ m/s^2$
Arm length	$0.201\ m$
Location of the center of mass of the pendulum	$0.154985\ m$
Moment of inertia of the arm	$0.0052\ kg.m^2$
Moment of inertia of the pendulum	$0.0023\ kg.m^2$
Pendulum length	$0.30997\ m$
Pendulum mass	$0.2866\ kg$

Was made an approximation in the calculation of the moments of inertia, both links are taken as constant circular bars and invariant mass

$$I_0 = \frac{1}{3} m_0 {L_0}^2 \tag{36}$$

$$I_1 = \frac{1}{12} m_1 {L_1}^2 \tag{37}$$

IOP Conf. Series: Journal of Physics: Conf. Series **850** (2017) 012005 doi:10.1088/1742-6596/850/1/012005

Where (36) is the moment of inertia of the arm; measured from the end connected to the motor shaft to the opposite end. While (37) is the pendulum moment of inertia; measured pendulum from the center of mass.

8) The way to export the data to Workspace of block diagram above is performed with the same aggregate block in step *6)*.

9) Taking the exported data in points *6)* and *8)* the following graphs were made.

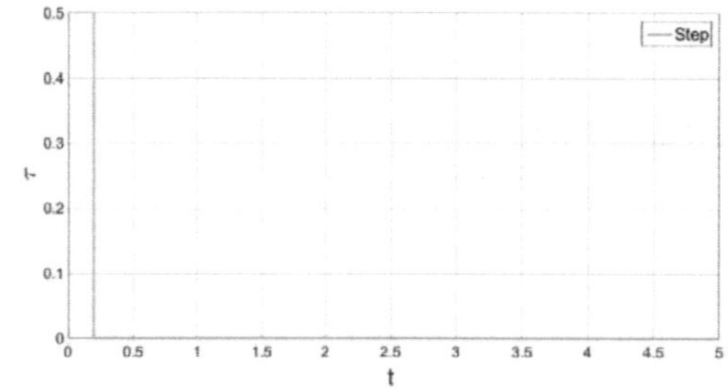

Figure 7. Step torque

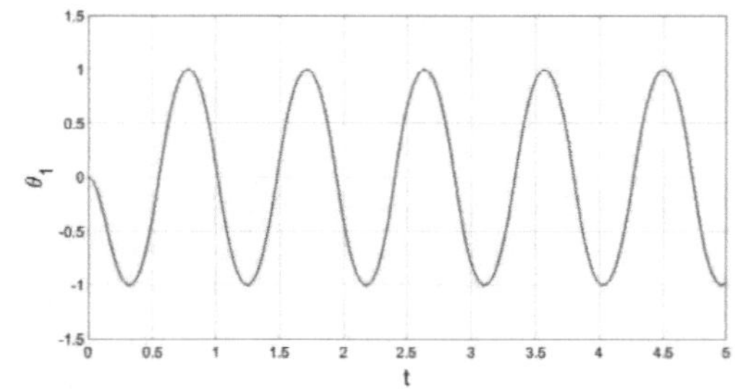

Figure 8. Angular position of pendulum with CAD model

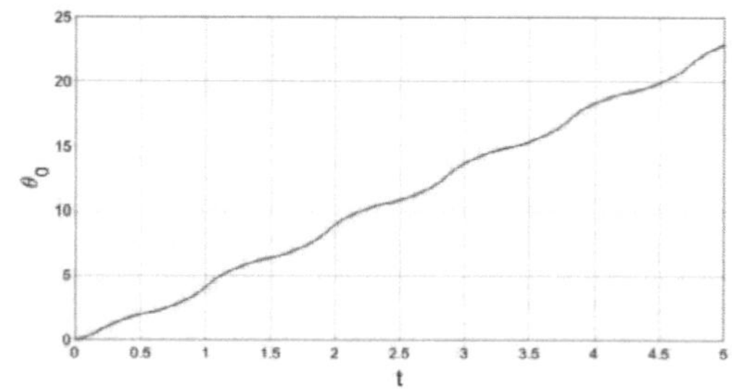

Figure 9. Angular position of arm with CAD model

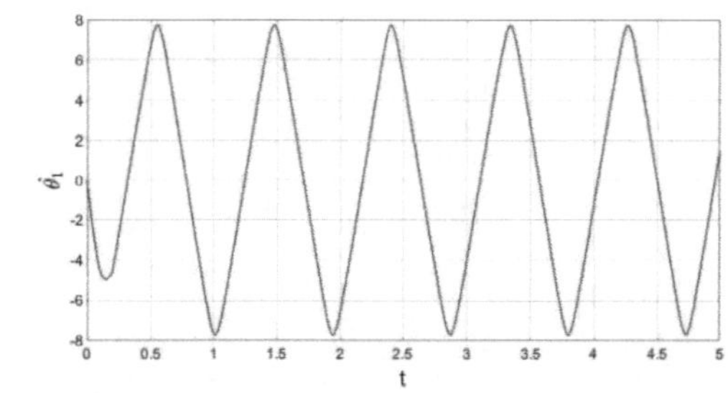

Figure 10. Angular velocity of pendulum with CAD model

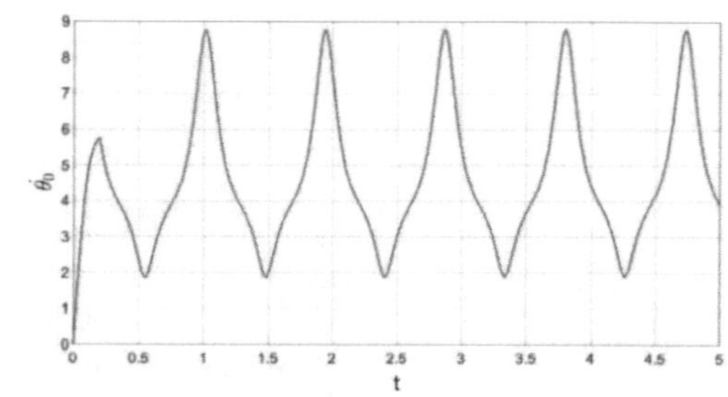

Figure 11. Angular velocity of arm with CAD model

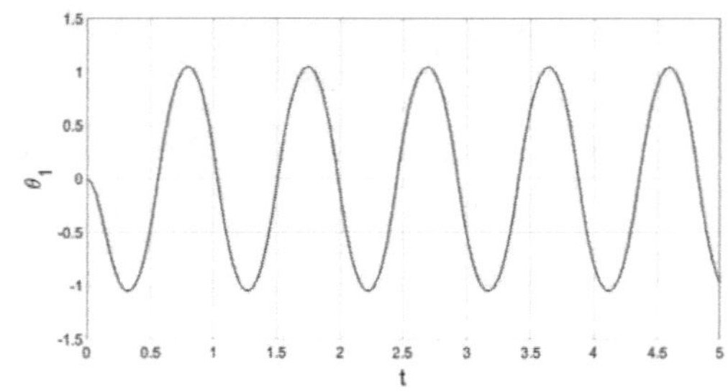

Figure 12. Angular position of pendulum with Euler-Lagrange's equations

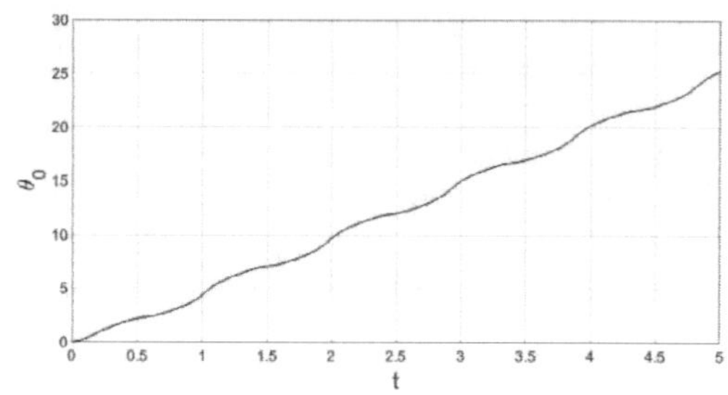

Figure 13. Angular position of arm with Euler-Lagrange's equations

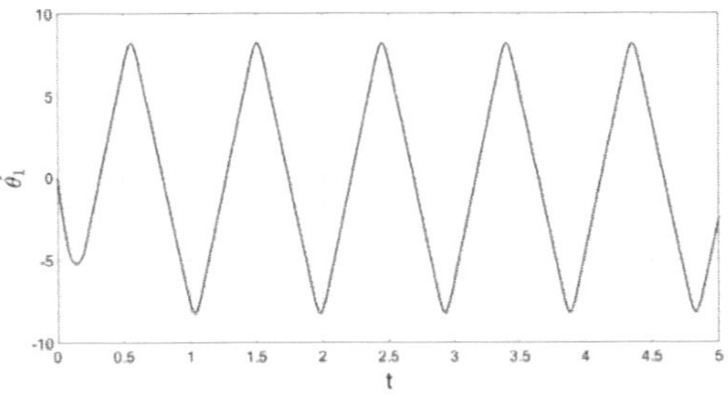

Figure 14. Angular velocity of pendulum with Euler-Lagrange's equations

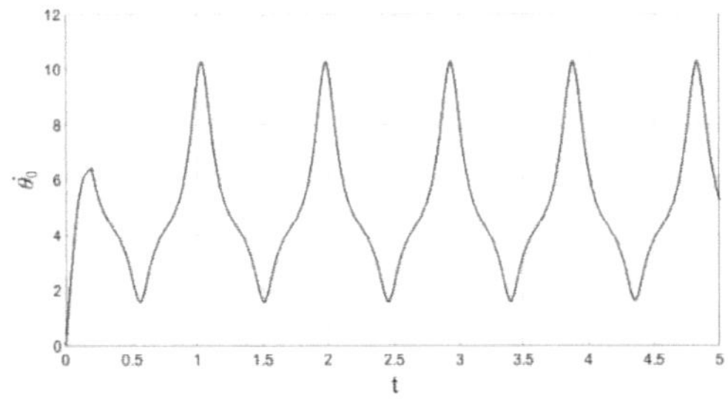

Figure 15. Angular velocity of arm with Euler-Lagrange's equations

It is noted that the system behavior is the same in both simulations. A superposition of results was performed by way of display the error between the CAD model simulated and the Euler-Lagrange equation, with the following results.

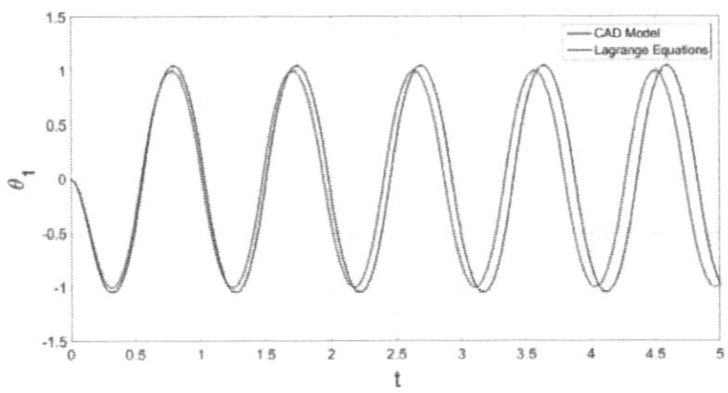

Figure 16. Comparison results angular position of pendulum

Figure 16 allows viewing behavior overlapping displacement of the center of mass in the pendulum. The red line represents the CAD model while the blue is the result of θ_1 graph the solution over time.

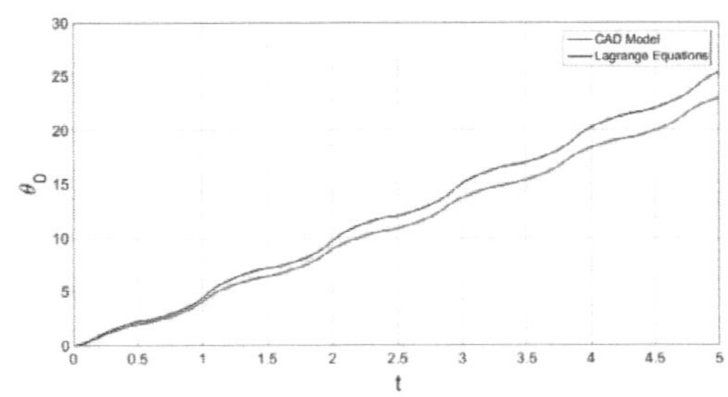

Figure 17. Comparison results angular position of arm

Figure 17 is overlaying the results to simulate and plot θ_0 function of time, where the red line is the result of simulating in SimMechanics the CAD model and the blue line is the solution of the Euler-Lagrange's equation.

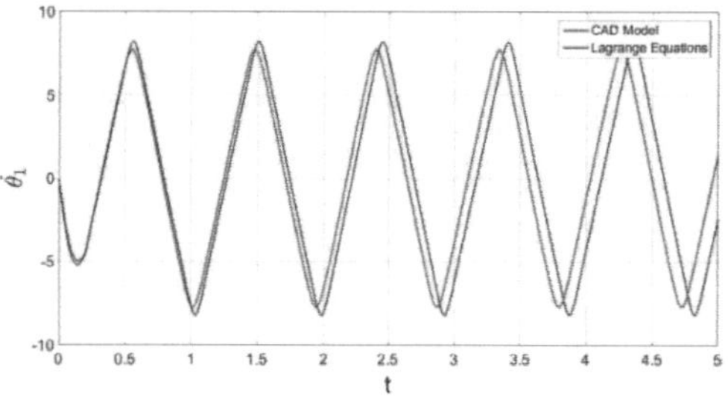

Figure 18. Comparison results angular velocity of pendulum

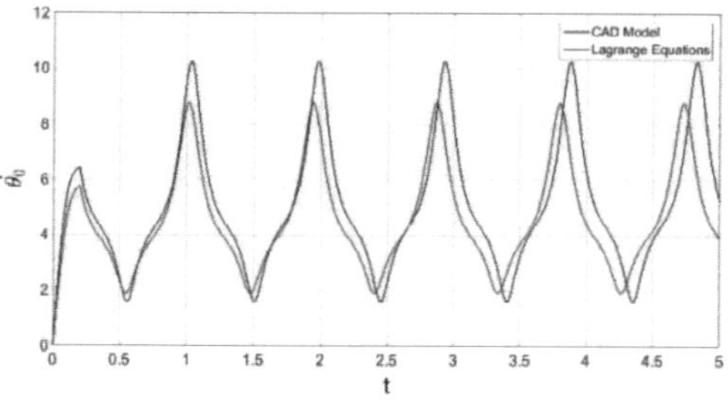

Figure 19. Comparison results angular velocity of arm

Similarly as with Figure 16 and 17, the Figure 18 and 19 illustrate the behavior of the angular velocity with respect to time for the pendulum and the arm respectively.

It is important to note that the graphs do not have a zero error due to the approximations that were made to the calculation of the moments of inertia of each link; the difference is that the SOLIDWORKS software performs a more accurate calculation of the moments of inertia, because the software considers the geometry of each link.

Then, the pendulum trajectory analysis shown, which part of the potential energy in the center of mass as shown in (9).

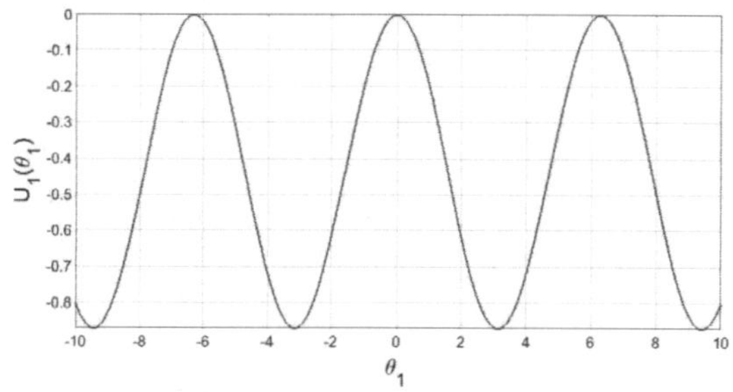

Figure 20. Potential energy

The phase space diagram is obtained when all the trajectories of the pendulum are gathered, which shows which are the critical points of the system. In this case, it's know that the pendulum has two critical points, a stable point at which the pendulum is in its obtained hanging position and unstable point in this case refers to a metastable point is satisfied when the pendulum is reversed. Graphically you can get the critical points in the system, we can first obtain which is the equation for calculating the minimum by Figure 20, it can be seen that the minimum occurs when the wave goes from a negative slope to a positive slope, therefore, we are obtained minimum $\pm\pi i$ where i must be odd. Similarly, one can find the maximum, these occur when moving from a positive slope to a negative, therefore the maximum are present in $\pm\pi i$ where i must be zero or even.

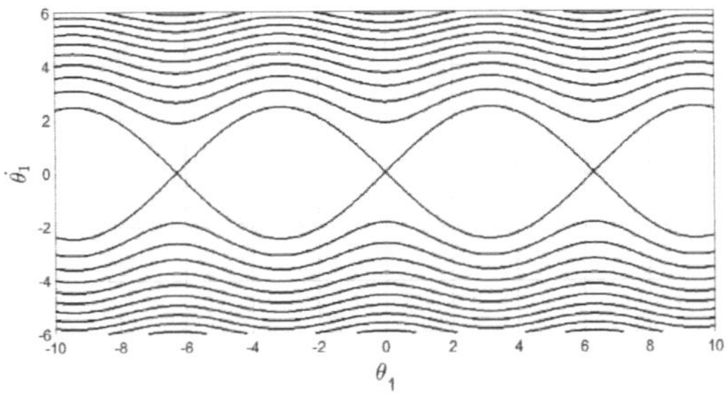

Figure 21. Pendulum phase space

Physically each point on the Figure 21 represents a possible system status, explicitly state in which the pendulum would.

As mentioned above the pendulum has two equilibrium points, however, in the phase space endless these points are. This is because there is shown the position like a straight or R space when the real space is a circle that is not topologically equals R. Therefore, if space representing geometric positioning system is not topologically equivalent to R, for subsequent phase space will not be in R^2, since this is the Cartesian product between the geometric space and space representing the speed is generally R. Topologically speaking the pendulum has a geometric space $S1$ which is equivalent to a circle and a space velocity R, therefore, the phase space $S1 \times R$ equivalent to a cylinder and not to R^2. The correct graphic process should be done on a cylinder [18] where its circumference is the geometric representation of the pendulum, this is equivalent to the set of possible positions of the center of mass, while the position along its axis represents the velocity of the center mass, this cylinder is topologically correct space phase. Touring the circumference of the cylinder we can see that there is only one minimum and one maximum diametrically opposed. In Figure 20 you could take energy levels were the result of drawing lines parallel to the axis of the θ_1, straight position, well, now those levels may be taken with circles surrounding the cylinder.

8. Conclusions

The formalism of Euler-Lagrange allows a dynamic modeling of the rotary inverted pendulum in a simple way, thanks to that the formalism allows working in a scalar manner. Checking one of its advantages that recommends using the method in case there are rotations in the system.

In this paper, the dynamic modeling of a rotary inverted pendulum has been presented step by step, contributing a complete document to the scientific community that wishes to work with this type of system. In addition, are shown simulations that they validate the equations of motion, comparing the above mentioned with the CAD model realized in SOLIDWORKS, showing valid results. In later studies, these simulations are a good point of reference.

This document differs from the others thanks to the use of the SimMechanics link, which allows multi-body simulations of simple and sophisticated mechanical systems, such as robots, underactuated systems or mechanisms of any kind. It can model sensors, actuators, joints, forces, etc. With the implementation of specialized blocks. Was successfully used this environment in the simulation of the proposed CAD model. Due to its developers it is possible be certain that it is a reliable tool. It is also, very simple to use; this is an advantage to perform simulations of 3D models, since it is not necessary to create platforms for their simulation and it is enough to make a CAD model in any software of design, in this case it was SOLIDWORKS. In addition, MATLAB allows to work with different models in the same environment, in this case, the Simulink-MATLAB environment, where there were realized the simulations of the mathematical model. While the SimMechanics-MATLAB environment was used to simulate the CAD model, it is necessary to clarify that SimMechanics is in the same environment Simulink, which generates comfort at the moment of realizing simulations, Since it is not necessary to change of software for each model.

Additionally, in this article in particular, the analysis of the trajectories of the pendulum is performed, the phase space shown is topologically correct, this being a good contribution to future dynamic, kinematic, chaotic and physical studies of the system.

References

[1] Ramírez C 2012 *Modelado y Análisis de Sistemas Mecánicos Subactuados* (San Luis Potosí: Instituto Potosino de Investigación Científica y Tecnología) p 213
[2] Acosta J 2010 Furuta's pendulum: A conservative nonlinear model for theory and practice *Mathematical Problems in Engineering* **2010** 484
[3] Cruz M, Silva R, Merlo C, Villarreal M, Muñoz D and Hernandez V 2014 *Int. Conf. On Mechatronics, Electronics and Automotive Engineering* (Cuernavaca: IEEE) vol 1 p 218
[4] Blajer W and Kolodziejczyk K 2008 Modeling of underactuated mechanical systems in partly specified motion *Journal of Theoretical and Applied Mechanics* **46** 1007
[5] Kajita S, Morisawa M, Miura K, Nakaoka S, Harada K, Kaneko K, Kanehiro F and Yokoi K 2010 IEEE/RSJ *Int. Conf. on Intelligent Robotics and Systems* (Taipei: IEEE) p 6239
[6] Younis W and Abdelati M 2009 *2nd Mediterranean Conf. on Intelligent Systems and Automation* (Zarzis: AIP Conference Proceedings) vol 1107 p 380
[7] Prieto A, Viguria A, Vela B, Fiacchini M and Cano R 2005 *PPCarro: Vehículo móvil basado en péndulo invertido* (Sevilla: Escuela Superior de Ingenieros de Sevilla) p 62
[8] Panomruttanarug B and Chotikunnan 2014 *11th Int Conf on Electrical Engineering/Electronics, Computer, Telecommunications and Information Technology* (Nakhon Ratchasima: IEEE) vol 1 p 501
[9] Careaga A, Katzschmann R and Almeida J 2013 *Furuta Pendulum* (Massachusetts: Massachusetts Institute of Technology) p 33
[10] Escobar L and Montoya O 2012 *Control Híbrido del Péndulo de Furuta* (Pereira: Universidad Tecnológica de Pereira) p 68
[11] Moreno E 2013 *Diseño, construcción y control de un péndulo de Furuta* (Tijuana: Instituto Politécnico Nacional) p 57
[12] Fedák V, Ďurovský F and Üveges R 2014 Analysis of Robotic System Motion in SimMechanics and MATLAB GUI Environment *MATLAB Applications for the Practical Engineer* (Rijeka :Intech) chapter 20 pp 565-581
[13] Meehan P 2016 Predicting and Simulation Chaotic Instabilities in and Inclined Furuta Pendulum *Chaotic Modeling and Simulation (CMSIM)* **1** 112
[14] Furuta K, Yamakita M and Kobayashi S 1991 *Int. Conf. on Industrial Electronics, Control and Instrumentation* (Kobe: IEEE ES/SICE) vol 3 p 2592
[15] Triana G, Buitrago J and Ramírez J 2015 Control de un péndulo de Furuta, una revisión del estado del arte *Scientia et Technica* **4** 97
[16] Cutberto C 2011 *Control Robusto H_∞ de Sensibilidad Mezclada Aplicado a Sistemas Lineales Invariantes en el Tiempo Subactuados* (San Nicolás de los Garza: Universidad Autónoma de Nuevo León) p 112
[17] Rafael K and Víctor S 2003 Modelo dinámico and Propiedades *Control de Movimiento de Robots Manipuladores* (Madrid: Pearson Educación S.A.) chapter 3-4 pp 59-119
[18] Gerving C 2013 *Dynamics of a Quantum Inverted Pendulum Explored in a SPIN-1 BEC* (Georgia: Georgia Institute of Technology) p 140

IOP Conf. Series: Journal of Physics: Conf. Series **850** (2017) 012006 doi:10.1088/1742-6596/850/1/012006

Effect of doping with Al/B on the sensitivity of a metallic carbon nanotube to CO_2

Aura Merlano[1,*], Andrés M. Garay-Tapia[2], F. R. Pérez[1] and Ángel Salazar[1]

[1]*Grupo de Óptica y Espectroscopía (GOE), Centro de Ciencia Básica, Universidad Pontificia Bolivariana, A.A. 56006 Medellín, Colombia.*
[2]*Centro de Investigación en Materiales Avanzados S.C. (CIMAV), Unidad Mty, Alianza Norte 202, Autopista Monterrey-Aeropuerto Km 10, Apodaca, N.L., México.*

[*]*Corresponding author: aura.merlano@upb.edu.co*

Abstract. In this work the effect of doping with aluminum (Al) and boron (B) an armchair (6,6) carbon nanotube on its sensibility to carbon dioxide (CO_2) for possible application in sensors of this gas was studied. Using first-principles calculations within the framework of the density functional theory (DFT), adsorption energies were obtained in the cases when the molecule is initially perpendicular to the surface of the nanotube, near the dopant atom, and located above a carbon atom of the nanotube, above a C-C bond, or directly above the center of a hexagon. It was found that doping with Al does not improve the adsorption of the molecule compared to the pristine nanotube. However, doping with B slightly favors the adsorption for some of the considered positions. The results suggest that B doping might be an acceptable option in the design and construction of nano devices for CO_2 detection.

1. Introduction

Liberation of carbon dioxide (CO_2) into the atmosphere due to the consumption of large amounts of fossil fuels as coal, petroleum and natural gas has become one of the most serious global environment problems, which is focusing attention of the scientific community[1-5]. Global warming concerns have triggered global efforts to reduce the concentration of atmospheric CO_2. Detection, capture and storage of CO_2 gas is considered a crucial strategy for reduction of anthropogenic emission [6].

The adsorption of CO_2 is very important in environmental and industrial applications. In earlier adsorption processes, the choice of an adsorbent was restricted to various types of activated carbons, silica and alumina. Nevertheless, emergence of novel materials has generated new theoretical approaches to adsorption phenomena. The main types of adsorbents of practical environmental importance are presented in Table 1 [7]. Particularly, there is growing interest for metal free carbon-based nanomaterials for gas adsorption. Carbon-based nanomaterials such as fullerene, carbon nanotubes and graphene offer superior thermal and chemical stability as CO_2 adsorbents [8].

Carbon nanotubes (CNTs) have excellent adsorption, catalytic and thermal properties and may be good candidates as CO_2 sensing due to their high conductivity, unique tubular structure, extraordinary chemical stability, and mechanical properties. Furthermore, they also have some advantages such as low power consumption, operation at room temperature, and size and weight reduction [9-14].

Though the adsorption capacity of CNTs remains limited, previous studies have demonstrated that the doping can tailor its electronic and structural properties to increasing its reactivity, enhance the range of molecular species to be detected and further improve the sensing performance [15],[16]. Consequently, the adsorption capability of CNTs can be improved through introducing heteroatom impurities (such as boron, aluminum, nitrogen, silicon, zinc, and titanium) and forming active sites in the nanotube walls [17]. Several researchers have theoretically studied platinum, aluminum, titanium and gold doped CNTs for detecting gas molecules such as CO, NO, NH_3, N_2, H_2 and C_2H_2 [18-19]. S. Ishii and Y. Takano [20] have reported that doping with alkali metals enhances the conductivity of graphite, resulting in superconductivity. Besides, diamond is superconductor by boron (B) doping. The CNT structure is intermediate between that of 2-dimensional graphite and 3-dimensional diamond. Considering these properties, researchers speculated that elemental B in CNTs would enhance their conductivity.

Now well, materials which simultaneously possess the unique properties of CNTs and high adsorption capability of aluminum (Al), or the unique properties of CNTs and excellent conductivity of B [21] could be useful for the CO_2 sensor technology with superior performance for gas detection.

In the present work, DFT calculations were conducted to investigate the effect of Al and B atoms doping on adsorption of CO_2 onto the surface of an armchair (6,6) CNT. Al possesses the same ns^2np^1 valence electric structure of B atom. We explore some possible energetically favorable sites of the nanotube for the adsorption of a CO_2 molecule perpendicular to the CNT surface.

The rest of the article is organized as follows. In Section 2 the details of the computational method based on DFT is presented. Section 3 shows the results related to adsorption energy and optimal positions of CO_2 on the surface of CNTs. Finally, in section 4, conclusions have been drawn on this research.

Table No. 1 Basic types of adsorbents used for environmental purposes [7].

Carbon adsorbents	**Mineral adsorbents**	**Other adsorbents**
Active carbons	Silica gel	Synthetic polymers
Activated carbon fibers	Activated alumina	Metalorganic microporous and mesoporous materials
Carbon Molecular sieves	Oxides of metals	Composite adsorbents
Carbonaceous nanomaterials	Zeolites	Mixed adsorbents

2. Computational Methods

The study is performed within DFT by using the Vienna ab initio simulation package (VASP) [22] under generalized gradient approximation (GGA) with Perdew-Burke-Ernzerhof (PBE) exchange and correlation. A plane wave basis set with kinetic energy cutoff of 600 eV is used and the Brillouin zone is sampled by adopting (1x1x10) k-points Monkhorst–Pack grid for all structures. Furthermore, a supercell geometry with periodic boundary conditions is used and a sufficiently large vacuum space is inserted between the neighbor nanotubes to avoid their interaction (15.07 Å).

Adsorption properties of CO_2 onto pristine armchair (6,6) CNT and Al and B doped nanotube are investigated. The diameter of the pristine nanotube is 8.19 Å and the length is 9.81 Å, corresponding to four unit cells. To generate an Al-doped CNT, a carbon atom is removed from the perfect hexagonal structure of the CNT and replaced by an Al atom (Fig. 1a). Similarly, we proceeded with the B-doped CNT (Fig. 1b).

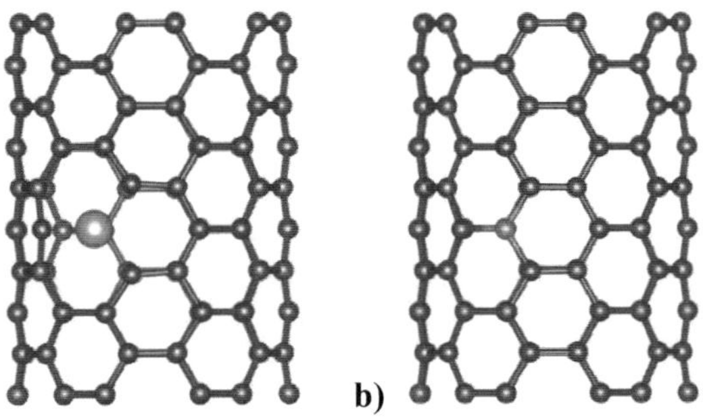

a) b)

Fig. 1 Optimized structure of (6,6) CNT with: **(a)** Aluminum (Al), **b)** Boron (B)

Different positions of the CO_2 molecule were considered. For pristine nanotube, three positions on the surface of the nanotube were considered: the molecule above of a carbon atom (Fig. 2a), above the center of the hexagonal ring (Fig. 2b), and above on the central point of a carbon-carbon bond (Fig. 2c). In each one of these positions, the CO_2 molecule was oriented perpendicular to the axis of CNT.

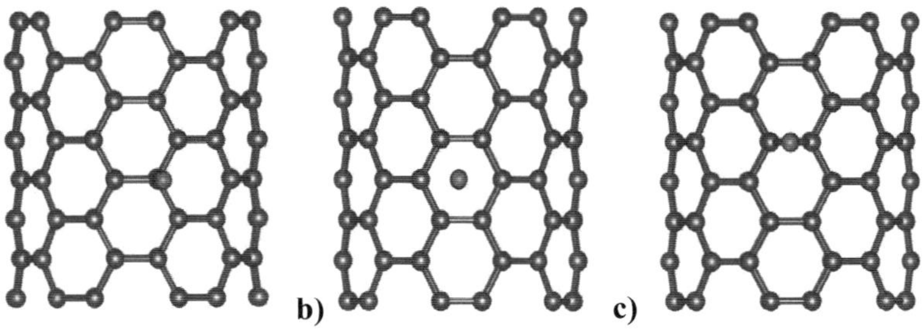

a) b) c)

Fig. 2 Positions of the CO_2 molecule on pristine nanotube (6,6).

We also considered different sites for the molecule on the Al-doped carbon nanotube: first, directly above the Al atom (Fig. 3a), second, above the center of a carbon atom ring including the aluminum atom (Fig. 3b), and third, above the central point of a carbon-carbon bond (Fig. 3c). The CO_2 molecule was also oriented perpendicular to the Al-doped CNT surface.

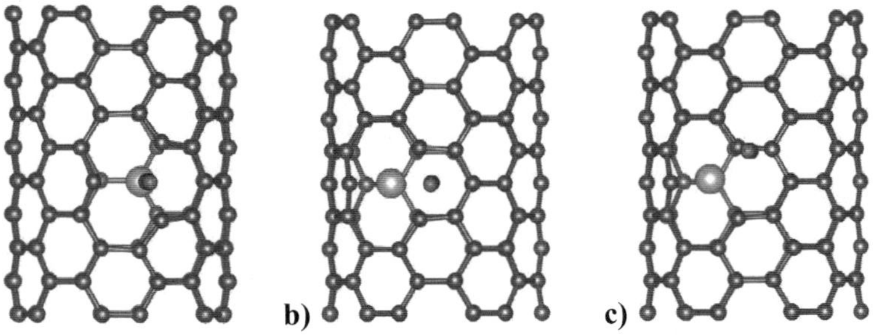

a) b) c)

Fig. 3 Positions of the CO_2 molecule on Al-doped (6,6) CNT.

Finally, for a B-doped CNT, positions of the gas molecule analogous to those studied in the Al-doped CNT were considered. First, the molecule directly above the B atom (Fig. 4a), second, the molecule above the center of a carbon atom ring including the boron atom (Fig. 4b), and third, above the central point of a carbon-carbon bond (Fig. 4c). The CO_2 molecule was also oriented perpendicular to the B-doped CNT surface.

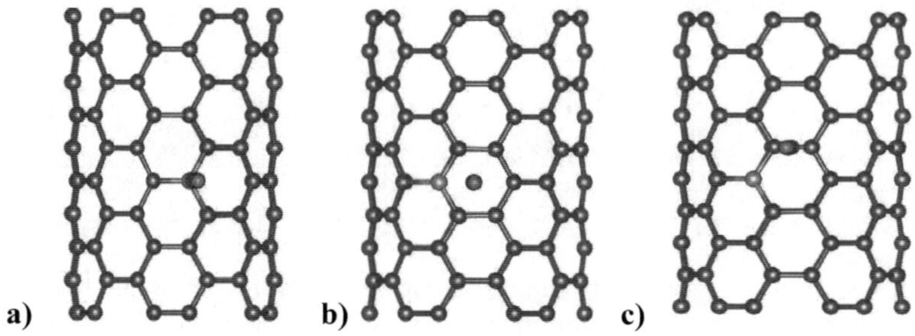

Fig. 4 Positions of the CO_2 molecule on B doped (6,6) CNT.

The adsorption energy E_{ads} of CO_2 onto nanotubes is defined according to Eq. (1), where $E_{tot}(CNT + CO_2)$ denotes the total energy of the nanotube-molecule system, and $E_{tot}(CNT)$ and $E_{tot}(CO_2)$ are the total energies of the isolated nanotube and gas molecule, respectively.

$$E_{ads} = E_{tot}(CNT + CO_2) - E_{tot}(CNT) - E_{tot}(CO_2) \quad (1)$$

3. Results and Discussion

Individual structural relaxations of pristine, Al- and B-doped CNTs and CO_2 molecule were performed. The optimized structures were obtained by relaxing each structure until the maximum force on any atom was not greater than 0.02 eV/Å. From these optimized structures, the nanotube-molecule systems for the different positions mentioned previously were constructed and new relaxations were performed for each one of the systems. Using Eq. (1), adsorption energies were calculated from the total energy of the respective system, and the isolated nanotube and gas molecule.

3.1. Adsorption of CO_2 on pristine (6,6) CNT

The calculated adsorption energies and final distances between molecule and pristine nanotube are shown in Table 2. In all cases, the adsorption energy is positive. The pristine nanotube and CO_2 above the center of the hexagonal ring and located perpendicular to the axial direction of the nanotube (Fig. 2b) corresponds to the configuration with minimum energy after relaxation (Figs. 5a and 5b). These results are in agreement with previous works where the adsorption of CO_2 on pristine CNTs for other positions was calculated too [9],[23-24].

Table No. 2 Results of adsorption energy for pristine CNT (6,6) with different positions of the CO_2 molecule.

Site	Orientation with respect to the CNT axis	E_{ads} (eV)	d (Å)	Figure
carbon atom	perpendicular	1.14	2.21	2a
hexagonal ring	perpendicular	0.62	2.46	2b
C-C bond	perpendicular	1.10	2.28	2c

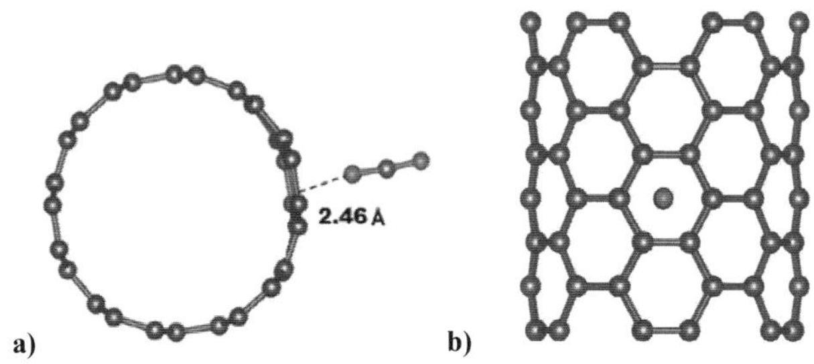

Fig. 5. Pure (6,6) CNT and CO_2 above the center of the hexagonal ring and located perpendicular to the axial direction of the nanotube. **(a)** Front and **(b)** side views of optimized structure.

3.2 *Adsorption of CO_2 on Al-doped (6,6) CNT*

Table 3 shows the results obtained for the Al-doped CNT and the different positions of the CO_2 molecule as presented above. The energetically most stable geometry occurs when the CO_2 molecule is initially on an C-C bond and located perpendicular to the axial direction of the nanotube (Fig. 6). However, all adsorption energies are higher compared to pristine nanotube. In this way, the interaction between the Al-doped CNT and CO_2 molecule is weak and doping with Al does not improve the adsorption of the CO_2 molecule.

Previous investigations have reported high adsorption capability of aluminium doped CNT. Wang *et al* [15] reported that Al-doped SWCNT presents high sensitivity to CO compared to pure SWCNT. Ganji *et al* [25] found that Pt adsorption ability of Al-CNTs is stronger than that of B and N doped CNTs. However, to the best of our knowledge, there is no report for adsorption of CO_2 on the surface of Al-doped (6,6) CNT. Then, it is a preliminary result showing a poor sensing ability of Al-doped (6,6) CNT to carbon dioxide and this motivates us to explore new powerful adsorbent for CO_2 molecule.

Table No. 3 Results of adsorption energy for Al-doped CNT (6,6) with different positions of the CO_2 molecule.

Site	Orientation with respect to the CNT axis	E_{ads} (eV)	d (Å)	Figure
Aluminum atom	Perpendicular	1.23	1.78	3a
hexagonal ring	Perpendicular	1.51	2.28	3b
C-C bond	Perpendicular	1.32	2.49	3c

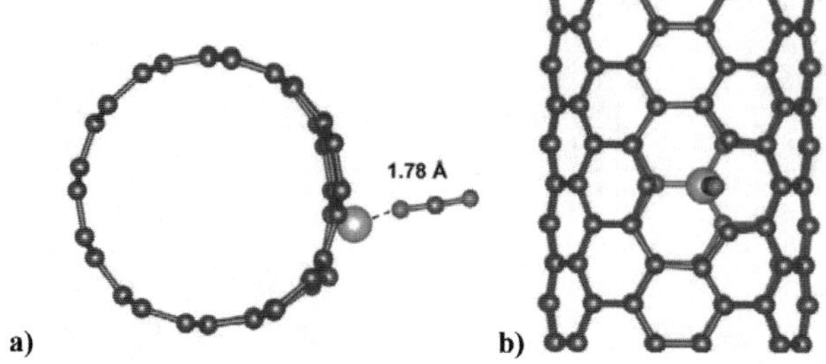

Fig. 6 Al-doped (6,6) CNT and CO_2 above on the central point of a Al-C bond and located perpendicular to the axial direction of the nanotube. **(a)** Front and **(b)** side views of optimized structure.

3.3 *Adsorption of CO_2 on B-doped (6,6) CNT*

Results for the B-doped (6,6) CNT are shown in Table 4. The energetically most favorable situation occurs when CO_2 is above the center of the hexagonal ring (Fig. 7). The B doping slightly favors adsorption of CO_2 compared to pristine and Al doped CNTs. It suggests that B doping may be an acceptable option to improve the CO_2 adsorption in nano devices for gas detection.. However, more studies have to be done providing a work direction to experimental scientists developing CNT-based chemical sensors.

Table No. 4 Results of adsorption energy for pristine B doped CNT (6,6) with different positions of the CO_2 molecule.

Site	Orientation with respect to the CNT axis	E_{ads} (eV)	d (Å)	Figure
boron atom	Perpendicular	1.07	2.10	4a
hexagonal ring	Perpendicular	1.04	2.55	4b
C-C bond	Perpendicular	1.05	2.42	4c

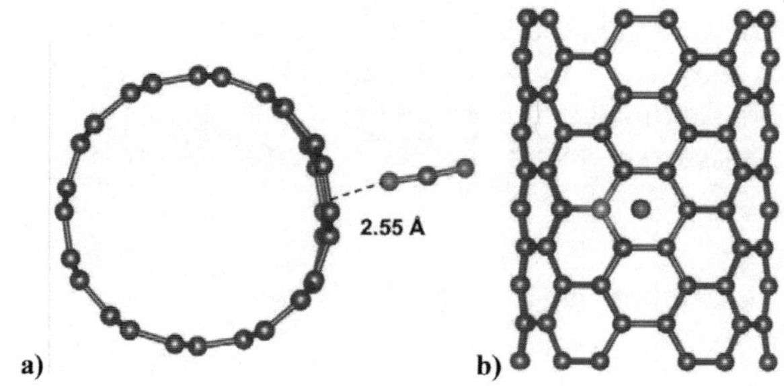

Fig. 7 B-doped (6,6) CNT and CO_2 above the center of the hexagonal ring and located perpendicular to the axial direction of the nanotube. **(a)** Front and **(b)** side views of optimized structure.

4. Conclusions

A first-principles DFT study on the adsorption of CO_2 gas molecule on pristine, Al and B-doped (6,6) armchair CNTs was performed. Results for pristine (6,6) CNT are consistent with those of other theoretical studies and suggest that adsorption capability for gaseous molecules is limited. Results also show that CO_2 adsorption ability of B-doped CNT is higher than that of an Al-doped CNT. Even so, adsorption is not considerably improved. However, B-doped CNT appears as a good candidate for detecting CO_2 molecules and more studies must be done. For future work and based on the obtained results, we could think of studying the influence of boron concentration in the adsorption properties of the nanotube to the CO_2.

5. Acknowledgments

Authors thank the financial support obtained from the Departamento Administrativo de Ciencia, Tecnología e Innovación, Colciencias, Colombia, through the call 617 for exchange of researchers Colombia-México UPB-CIMAV, contract 684 UPB-Colciencias. A. S. Merlano thanks Colciencias the financial support received under Convenio 0208-2014 UPB-Colciencias-Convocatoria 645 Nacional Jóvenes Investigadores e Innovadores 2014.

References

[1] International Energy Agency, "CO_2 Emissions From Fuel Combustion Highlights," 2013.
[2] U. EPA, "US Environmental Protection Agency." .
[3] U. S. Epa, "Inventory of U . S . Greenhouse Gas Emissions and Sinks," 2015.
[4] P. N. Jiří Čejka, Naděžda fIlková, *Molecular Sieves: From Basic Research to Industrial Applications*. Prague, Czech Republic, 2005.
[5] S. U. Pamela A. Matson (Chair), "Advancing the Science of Climate Change," in *NAS - America's Climate Choices*, 2010, p. 2.
[6] D. Y. C. Leung, G. Caramanna, and M. M. Maroto-Valer, "An overview of current status of carbon dioxide capture and storage technologies," *Renew. Sustain. Energy Rev.*, vol. 39, pp. 426–443, 2014.
[7] I. Dekany, *Adsorption and Nanostructures*. Springer-Verlag Berlin Heidelberg, 2002.
[8] S. W. de Silva, A. Du, W. Senadeera, and Y. Gu, "Neutral and charged boron-doped fullerenes for CO_2 adsorption," *Beilstein J. Nanotechnol.*, vol. 5, no. 1, pp. 413–418, Apr. 2014.
[9] A. Srirangarajan and M. U. Kahaly, "Ab Initio Study of Topological Defects in Single Walled Carbon Nanotubes and their Effect on Gas Sensing Mechanism," *Enabling Sci. Nanotechnol.*, vol. 1341, pp. 379–383, 2011.
[10] E. C. Jesús, J. Li, and C. R. Cabrera, "Latest Advances in Modified / Functionalized Carbon Nanotube-Based Gas Sensors," no. 2, 2013.
[11] P. Bondavalli, P. Legagneux, and D. Pribat, "Carbon nanotubes based transistors as gas sensors: State of the art and critical review," *Sensors Actuators, B Chem.*, vol. 140, pp. 304–318, 2009.
[12] N. Peng and Q. Zhang, "Sensing mechanisms of carbon nanotube based NH_3 gas detectors," pp. 1–15, 2002.
[13] E. Akbari *et al.*, "Analytical calculation of sensing parameters on carbon nanotube based gas sensors.," *Sensors (Basel).*, vol. 14, pp. 5502–15, 2014.
[14] Q. Wang and B. Arash, "A review on applications of carbon nanotubes and graphenes as nano-resonator sensors," *Comput. Mater. Sci.*, vol. 82, pp. 350–360, 2014.
[15] R. Wang, D. Zhang, W. Sun, Z. Han, and C. Liu, "A novel aluminum-doped carbon nanotubes sensor for carbon monoxide," *J. Mol. Struct. THEOCHEM*, vol. 806, no. 1–3, pp. 93–97, 2007.
[16] P. Shao, X. Y. Kuang, L. P. Ding, J. Yang, and M. M. Zhong, "Can CO_2 molecule adsorb effectively on Al-doped boron nitride single walled nanotube?," *Appl. Surf. Sci.*, vol. 285, no. PARTB, pp. 350–356, 2013.
[17] P. A. Gowri Sankar and K. Udhayakumar, "Electronic properties of boron and silicon doped (10, 0) zigzag single-walled carbon nanotube upon gas molecular adsorption: A DFT comparative study," *J. Nanomater.*, vol. 2013, no. 2, 2013.
[18] C. S. Yeung, L. V. Liu, and Y. A. Wang, "Adsorption of Small Gas Molecules onto Pt-Doped Single-Walled Carbon Nanotubes," *J. Phys. Chem. C*, vol. 112, no. 19, pp. 7401–7411, 2008.

[19] D. R. Kauffman, D. C. Sorescu, D. P. Schofield, B. L. Allen, K. D. Jordan, and A. Star, "Understanding the sensor response of metal-decorated carbon nanotubes," *Nano Lett.*, vol. 10, no. 3, pp. 958–963, 2010.

[20] S. Ishii and Y. Takano, "High-conductivity boron-doped carbon nanotubes," *SPIE Newsroom*, p. 2, 2007.

[21] D. Kang, X. Yu, M. Ge, F. Xiao, and H. Xu, "Novel Al-doped carbon nanotubes with adsorption and coagulation promotion for organic pollutant removal," *J. Environ. Sci.*, pp. 1–12, 2016.

[22] "The VASP site." [Online]. Available: http://www.vasp.at/. [Accessed: 05-Oct-2016].

[23] J. T. W. Yeow and Y. Wang, "A review of carbon nanotubes-based gas sensors," *J. Sensors*, vol. 2009, 2009.

[24] A. a. EL-Barbary, G. H. Ismail, and A. M. Babeer, "Effect of Monovacancy Defects on Adsorbing of CO, CO_2, NO and NO_2 on Carbon Nanotubes: First Principle Calculations," *J. Surf. Eng. Mater. Adv. Technol.*, vol. 3, no. 2, pp. 287–294, 2013.

[25] M. D. Ganji, M. G. Ahangari, and A. Khosravi, "Doping of carbon nanotubes with aluminum atom to improve Pt adsorption," *Appl. Surf. Sci.*, vol. 290, pp. 86–91, 2014.

Electrostatic plasma simulation by Particle-In-Cell method using ANACONDA package

J S Blandón, J P Grisales and H Riascos

Plasma, Laser and Applications Group, Universidad Tecnologica de Pereira, Pereira, Colombia

E-mail: jsblandon@utp.edu.co jpgrisales@utp.edu.co hriascos@utp.edu.co

Abstract. Electrostatic plasma is the most representative and basic case in plasma physics field. One of its main characteristics is its ideal behavior, since it is assumed be in thermal equilibrium state. Through this assumption, it is possible to study various complex phenomena such as plasma oscillations, waves, instabilities or damping. Likewise, computational simulation of this specific plasma is the first step to analyze physics mechanisms on plasmas, which are not at equilibrium state, and hence plasma is not ideal. Particle-In-Cell (PIC) method is widely used because of its precision for this kind of cases. This work, presents PIC method implementation to simulate electrostatic plasma by Python, using ANACONDA packages. The code has been corroborated comparing previous theoretical results for three specific phenomena in cold plasmas: oscillations, Two-Stream instability (TSI) and Landau Damping(LD). Finally, parameters and results are discussed.

1. Introduction

Plasma physics is currently an important discipline since 99% of matter in the universe is plasma [1]. In fact; nuclear fusion, telecommunications, nanotechnology, laser ablation, and others, requires accurate theoretical and experimental developments to explain physical mechanisms that are not fully understood [1]. In this paper, the plasma studied is electrostatic, it means, its behavior is ideal and representative since it allow to explain phenomena such as plasma oscillations, instabilities and damping.

Plasma electrostatic simulation has an important role in the development of plasma physics since is a useful tool to validate theoretical models, which being put into practice, would demand expensive or even not developed technologies. Birdsall and Langdon put the basis for the plasma simulation in their book "Plasma physics via computer simulation" [2]; which provides the necessary background for construction of codes that can represent oscillations, waves, instabilities and Landau damping within plasma. Morse and Nielson in [3] simulate TSI for 1-D, 2-D and 3-D dimensions. However, computational tools have evolved allowing more accurate simulations and efficient use of computational resources like the code develop by Martin n[4] who simulated in C, plasma oscillations, two stream and four stream instabilities in an electrostatic cold plasma.

The present work is a simulation of electrostatic plasma made in python using ANACONDA [5] libraries and is an implementation of the PIC algorithm, a common technique to simulate the motion of charged particles, which consist in the creation of a fixed mesh to calculate electromagnetic fields and a variable mesh to compute velocities and positions, which are updated every step for each particle. The Code was corroborated through analysis of characteristic phenomena of electrostatic cold plasma such as oscillations, TSI and LD. This was achieved by means of phase space graphics to observe the behavior of such phenomena over time, and dispersion relation (DR) graphic to compare simulated and theoretical data.

2. Theoretical Fundation

Plasma is electrostatic when the magnetic field $\vec{B}$ is equal to zero, and it is cold when particles temperatures is near to zero ($T_{i,e} = 0$) [6]. Therefore, the equations used in this model are *Gauss law* and *Lorentz Force*:

$$\nabla \cdot \vec{E} = \frac{\rho}{\epsilon_0} \tag{1}$$

$$\frac{d\vec{v}}{dt} = \frac{q\vec{E}}{m} \tag{2}$$

In equation (1), $\vec{E}$ corresponds to the electric field, ρ to the charge density, and ϵ_0 to electric permittivity in the vacuum. For equation (2), $\vec{v}$, q and m are the charge, mass, and velocity respectively.

For cold plasma, initial velocities are zero ($v = 0$), but for TSI them corresponds to a Maxwell distribution:

$$f(x,v) = \frac{n_o}{2}\left(\frac{1}{\sqrt{2\pi}v_{th}}e^{\frac{-(v-v_h)^2}{2v_{th}^2}} + \frac{1}{\sqrt{2\pi}v_{th}}e^{\frac{-(v+v_h)^2}{2v_{th}^2}}\right) \tag{3}$$

Where v is particle velocity, vh beam's average velocity, v_{th} is the thermal velocity, and n_0 is the particles reference density.

LD effect was also consider. This occurs when there is an energy exchange between a wave and particles in plasma with their velocity approximately equal to phase velocity of the wave.

3. Implementation and simulation: PE1D.py

Implementation of an electrostatic model was made through a python code named *Plasma Electrostático 1D (PE1D.py)* using ANACONDA packages. It requires a discretization and a normalization of involved equations in order to be included in a computational loop called PIC. Also, an initial setup is needed in order to start the simulation. After its run, PE1D.py shows graphics which allows understanding how the computational system works. Construction of PE1D.py is based on codes: espic.py by Gibbon [7], 1D PIC code by Fitzpatrick [8] and Electrostatic 1D Particle In Cell in Matlab/Octave of Markidis [9].

3.1. PIC Algorithm

PIC algorithm is usually used to simulate particle dynamics. This loop requires a bunch of conditions that the coder must include. Figure 1 shows a diagram of the method.

For electrostatic case, PIC algorithm comprises the following steps:

(i) Movement equation solutions for each particle and actualization of Lagrangian mesh.
(ii) Calculus of plasma charge density through assignation of particles to a Eulerian mesh, across particles were moving .
(iii) Computation of the electric field (E_x) through the solution of equation 1.
(iv) Electric field (E_x) is passed to particles making them move. Then return to step 1.

This algorithm, as was said before, it is a loop which requires a previous background such as an initial setup, a discretization and normalization of model equations.

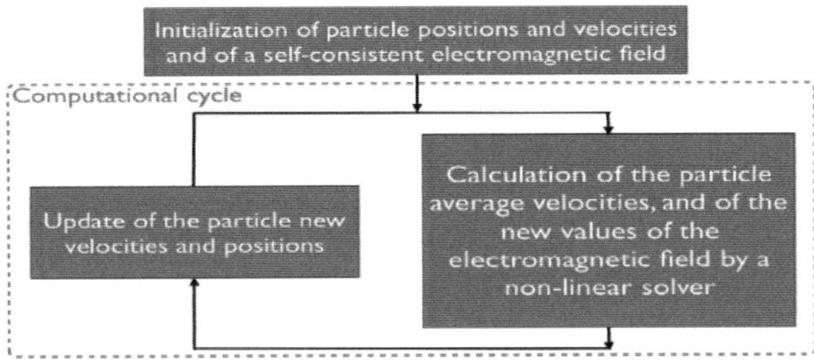

Figure 1. PIC algorithm [10].

3.2. Initial Setup

Initial setup consist of creation of superparticles that emulates plasma as well as space trough particles move. Therefore, a mobile mesh anchored to particles (Lagrangian) and a fixed mesh (Eulerian) are created [11]. To ensure the number of particles remains constant, periodic boundary conditions were added to the code, as made by Gibbon[7]. To finish initial setup, a first iteration needed by PIC its made.

Code PE1D.py was setting considering different phenomena. For cold plasma oscillations, system was built using 2048 particles, 256 grid, a 4π grid length points, 150 time steps, a $dt = 0.1$, with a sinusoidal spacial perturbation of 0.001 of amplitude as was proposed by Birdsall and Langdon [2]. Instead, for TSI time variables still, but 20000 particles, 1024 grid points, 32π grid length, 2.5 average beam velocity, and a sinusoidal velocity perturbation were used. Finally, for LD system setup remains the same, but with 1200 time steps and without perturbation.

3.3. Discretization

For electrostatic model implementation using PIC, equations 1 and 2 must be discretized:

$$x_i^{n+1} = x_i^n + v_i^{n+\frac{1}{2}}\Delta t \tag{4}$$

$$v_i^{n+\frac{1}{2}} = v_i^{n-\frac{1}{2}} + \frac{q_i}{m_i}E_i^n\Delta t \tag{5}$$

Equations 4 and 5 correspond to the Leapfrog Scheme [12]. Density is written in equation 6 using the Cloud In Cell (CIC) weighting scheme describe in [4].

$$\rho_j^{n+1} = \sum_i q_i\left(1 - \frac{|x_i - x_j|}{\Delta x}\right) \tag{6}$$

The electric field is written using the finite differences method, specifically using central difference form which allows a better approximation of equation 2.

$$E_{j+\frac{1}{2}}^{n+1} = E_{j-\frac{1}{2}}^{n+1} + \rho_j^{n+1}\Delta x \tag{7}$$

To assign electric field to particles, it is convenient to use the same weighing scheme as the used in density computation[4].

3.4. Normalization

Normalized units are important in computational physics because saves computational resources since quantities involved do not have many decimals [13]. Also, normalization of physical quantities allows use of dimensionless equations, enabling code transparency [7].

Simulated quantities for PE1D.py are as follows:

$$\omega_p t \to t \quad \omega_p x/c \to x \quad v/c \to v \tag{8}$$

$$eE/m\omega_p c \to E \qquad n_{e,i}/n_0 \to n_{e,i} \tag{9}$$

where t is time, w_p is plasma frequency, x is position, v is velocity, c is light velocity, E is electric field, m is mass, and n is particle density.

According to 8 and 9, the following quantities are considered as unitary: q_e, m_e, n_0.

3.5. Simulation

PE1D.py satisfies electrostatic model because PIC algorithm ensures that particles responds to the field which is created by themselves, this property is known as collective behavior and it is characteristic of every plasma. Here we considered electric field only.

The parameters of simulation are selected according to conditions proposed by Forslund [14]:

$$\omega_p \delta t < 2 \tag{10}$$

$$\lambda_d \geq \frac{\delta x}{2} \tag{11}$$

w_p was normalized, and λ_d (Debye length) represents the length of a cell in the mesh. They are related according with the relation deduced by Lindman [15]:

$$\delta t \leq \frac{2\pi}{\omega_p + \frac{\pi v_{th}}{\delta x}} \tag{12}$$

Again $v_{th} = 1$ for PE1D case. Relation 12 is important because ensure reduction of computational noise of the system which drives to decrease of non physical effects. Moreover, electric field is calculated directly from Gauss equation without computation of electric potential,so this simplifies the model and avoids adding computational noise from an extra step.

Once system is perturbed, the amplitude of it must be small respect to the equilibrium quantities. In TSI case, means perturbed velocity is lesser than average beam velocity(v_h) [1].

For TSI, theoretical DR where built according to code by Markidis [16]. On the other hand, for simulated DR the two dimensional Fast Fourier Transform algorithm (FFT) was used in spatial and temporal components of electric field [10], [17]. In addition, electric field has a magnitude of number of cells,so this number was conveniently chosen as a power of two, having into account the nature of the FFT.

To visualize PE1D.py conservative behavior it were plot energy in figure 3, considering potential, kinetic, and total energy.

4. Results and Discussion

4.1. Cold plasma oscillations

Cold plasma oscillations were simulated. Figure 2 shows electric field in the Eulerian Mesh. It corresponds to time step 150, and displays an oscillatory behavior as predicted by the cold plasma oscillations theory. Sinusoidal particle distribution is shown in density and hence in electric field generated by the particles, thus collective behavior of plasma is remarked.

Birdsall and Langdon [2] explains that under a sinusoidal perturbation, the system behaves like an harmonic oscillator, showing an exchange between kinetic and potential energy. Figure

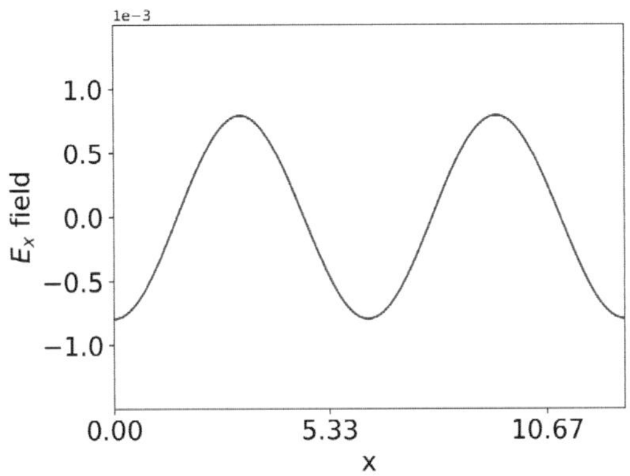

Figure 2. Electric field for cold plasma oscillations.

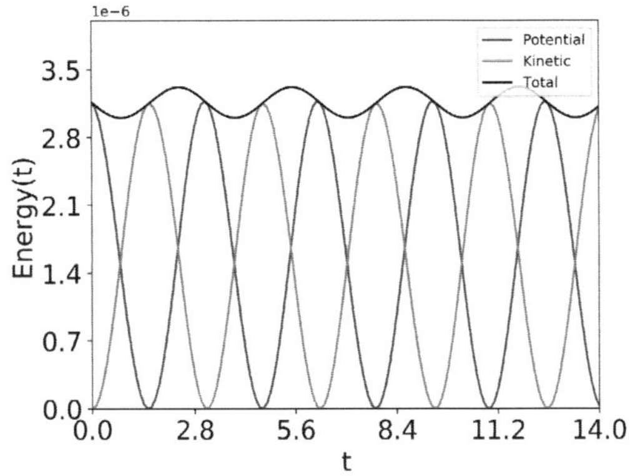

Figure 3. Energies as a function of time for cold plasma oscillations.

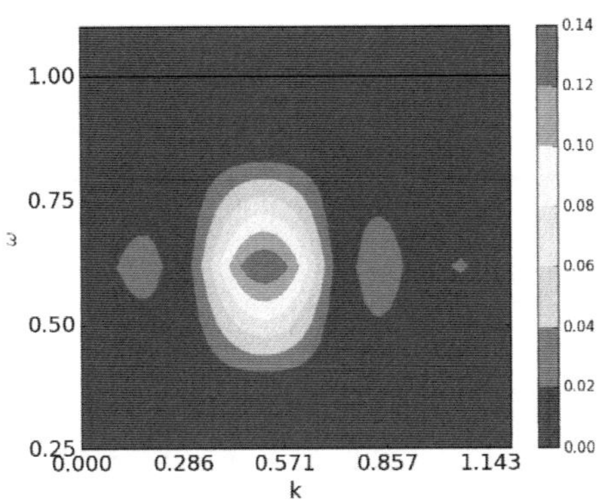

Figure 4. Dispersion relation for cold plasma oscillation: theoretical and simulated. The black line corresponds to theoretical DR. Different colors means different particles densities which oscillates in such frequency. A higher density is symbolized through warm colors such as red, orange and yellow. Instead, lower density is represented by colors such as blue and green. It is clear from the image that most of particles oscillates at $\omega = 0.6$, i.e near to plasma frequency as expected by theory [1].

3 shows energy system vs time, shows an oscillatory behavior according with the theory predictions.

Figure 4 is the DR and displays the temporal and spatial frequencies one against another for cold plasma oscillations.

4.2. Two-Stream instability

For TSI considerations were take into account: first, the temporal evolution in the phase space to corroborate if it presents its characteristic vorticity [1], and second respective dispersion relation to check the correct simulation of the field. Three test were made with 150 time steps and amplitudes of velocity perturbation of 0, 0.1 and 0.5 for figures 5,6,7 respectively.

Figures 5,6 and 7 are phase spaces plots for TSI with different amplitude perturbation. Those results corresponds to two identical electron flows in a one dimensional plasma. Such case was studied by Fitzpatrick [8]. Results obtained by PE1D.py are according with [8]. Likewise, it is noted a characteristic vorticity of this instability as shows figures 5(b), 6(b) and 7(b) [3]. In figure 5(a), instability appears without external perturbation, it means that it appears only because of particle movement. Clearly is observe that particles suffer a heating because movement as shown by Martin [4]. On the other hand, figures 6(a) and 7(a) instability shows different states of development, which suggest a relation between amplitude perturbation and instability emergence.

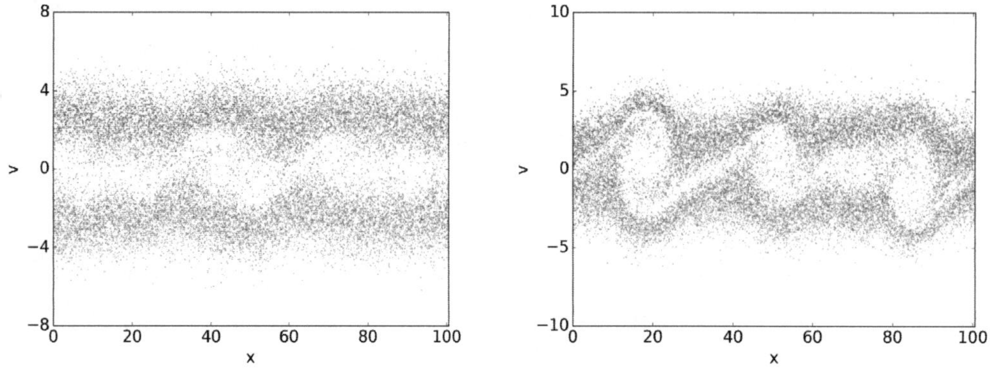

(a) t=100 (b) t=150

Figure 5. Phase space for TS instability in cold plasma: no perturbed.

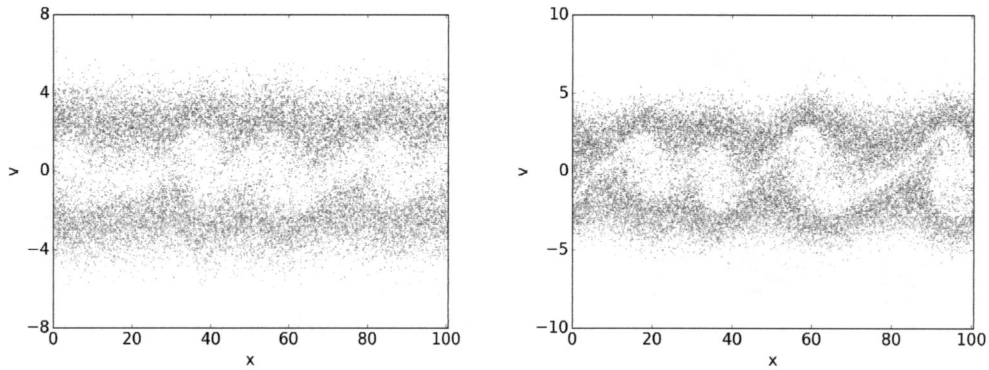

(a) t = 100 (b) t = 150

Figure 6. Phase space for TS instability in cold plasma: 0.1 amplitude of perturbation.

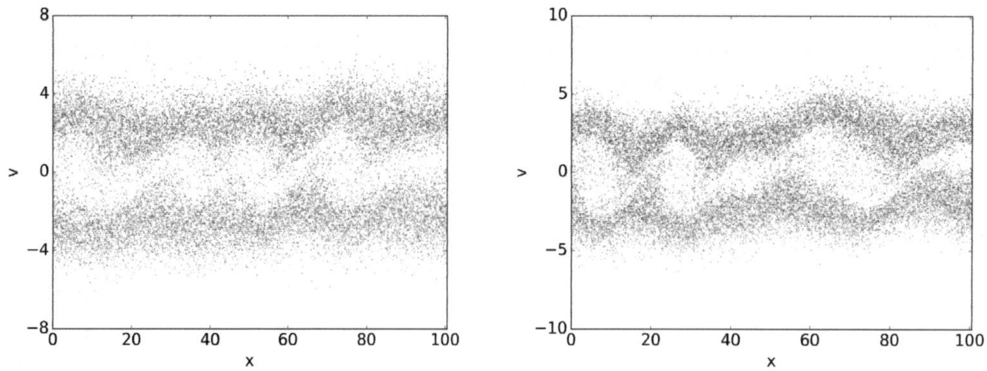

(a) t = 100 (b) t = 150

Figure 7. Phase space for TS instability in cold plasma: 0.5 amplitude of perturbation.

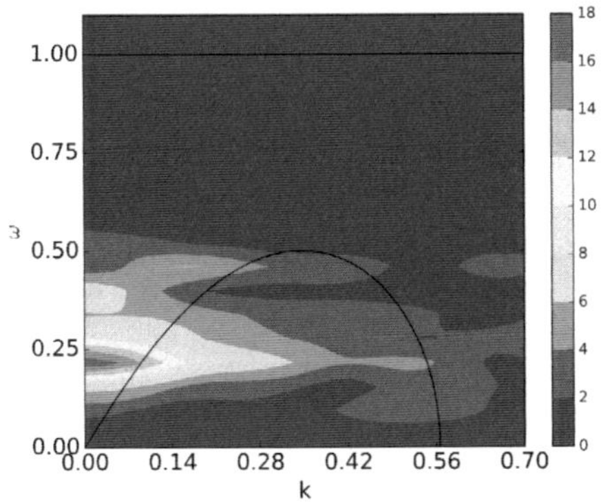

Figure 8. Dispersion relation for TS instability in cold plasma: theoretical and simulated without perturbation.

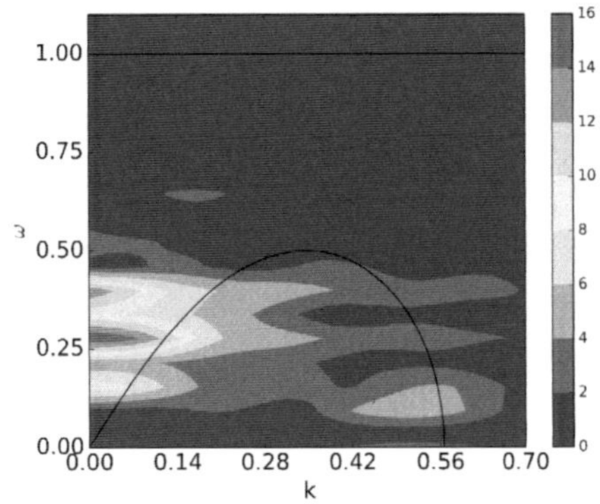

Figure 9. Dispersion relation for TS instability in cold plasma: theoretical and simulated with 0.1 amplitude of perturbation.

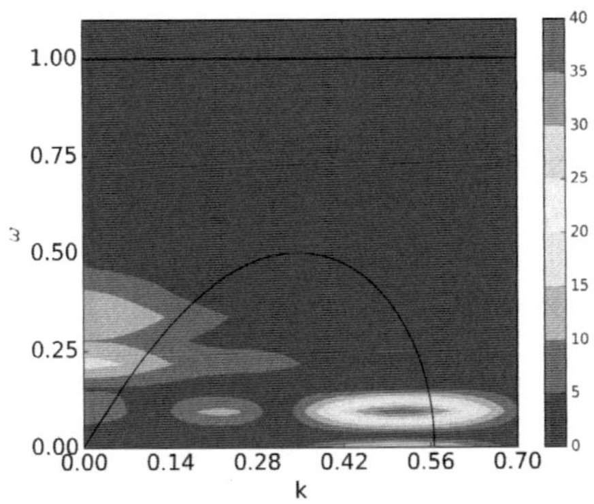

Figure 10. Dispersion relation for TS instability in cold plasma: theoretical and simulated with 0.5 amplitude of perturbation.

About DR in figures 8,9, and 10, the black line on top is plasma frecuency and the black curve is theoretical DR [2]. Figure 8 shows that particles oscillate near the first wavelengths in theoretical DR, while in figure 6 oscillations are better distributed along it, possibly as an effect of perturbation. In figure 10 with a perturbation near to average beam velocity, particles oscillates near to larger wavelengths in theoretical DR.

4.3. Landau damping

PE1D.py allows to simulate LD. Results obtained appears after a TSI simulation. As was mentioned in section 3.2, system has the same setup as TSI with time step at 1200 as shown in 11(a). There is a evidence of a particle trapping as Birdsall and Langdon exposed [2]. Figure 11(b) is the $\vec{E}$ field energy against time. The results obtained in this work are in agreement with [2]. Thus, previous results confirm the validity of electrostatic cold plasma simulation .

5. Conclusions

According to the results, it is conclude that perturbation amplitude accelerates emergence of TS instability. Moreover, perturbation causes that particles oscillating underneath plasma frequency tend to settle according to theoretical DR. Finally, results from different simulated phenomena such as cold plasma oscillations, TS, and LD, had corroborated the validity of PE1D.py as electrostatic unidimensional plasma computational representation.

IOP Conf. Series: Journal of Physics: Conf. Series **850** (2017) 012007 doi:10.1088/1742-6596/850/1/012007

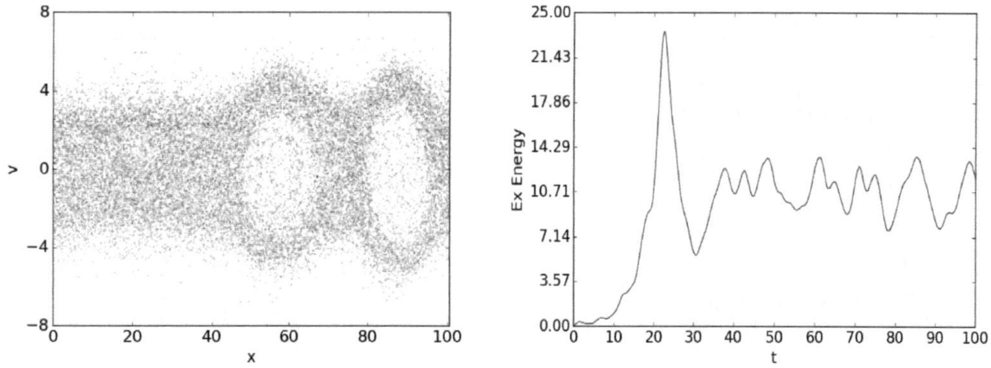

(a) Phase space (b) $\vec{E}$ Field Energy

Figure 11. Phase space (a) with t=1200 and $\vec{E}$ Field Energy (b) showing LD for TS instability.

Acknowledgments

We would like to thanks to Henry Riascos Landázuri, director of the Plasma, Laser and Applications Group(GPLA) and Universidad Tecnológica de Pereira (UTP), for his support to develop this work. Also, we acknowledge the resources given by SENA and Colciencias to do this simulation project.

References

[1] Chen Francis F 1974 *Introduction to plasma physics* (Springer, New York)
[2] Birdsall C K and Langdon A B 1984 *Plasma physics via computer simulation* (McGraw-Hill)
[3] Morse R L and Nielson C 1969 *Physical Review Letters* **23** 1087
[4] Martin D 2007 *Electrostatic PIC simulation of plasmas in one dimension* (Citeseer)
[5] Analytics C 2016 Anaconda software distribution URL `https://contiuum.io`
[6] Guasp J 1976 *Elementos de teoria cinética de plasmas* (MALLORCA. CONSEJO INSULAR)
[7] Gibbon P 2014 *ESPIC.py* (CERN)
[8] Fitzpatrick R 2006 *Computational physics: An introductory course* (University of Texas)
[9] Markidis S 2016 *pic.m* (PDC Center For High Performance Computing (KTH))
[10] Markidis S and Lapenta G 2011 *Journal of Computational Physics* **230** 7037–7052
[11] Woolfson M M and Pert G J 1999 *An introduction to computer simulation* (Oxford University Press,)
[12] McMillan S 2013 *Computational Physics*
[13] Lapenta G 2006 *Centrum voor Plasma Astrofysica Katholieke Universiteit Leuven*
[14] Forslund D W 1985 *Space Science Reviews* **42** 3–16
[15] Lindman E 1970 *Journal of Computational Physics* **5** 13–22
[16] Markidis S 2014 *dispersion.m* (PDC Center For High Performance Computing (KTH))
[17] Matsumoto H and Omura Y 1985 *Computer simulations of space plasmas*

IOP Conf. Series: Journal of Physics: Conf. Series **850** (2017) 012008 doi:10.1088/1742-6596/850/1/012008

Evaluation of the activity of a gold catalyst supported on alumina (Au/γ–Al_2O_3)

M Barrera [1], R Gómez [1], C Caballero[1], J Oviedo [2] and L C Sánchez[3]

[1] Departamento de Química, Universidad de Córdoba, Carrera 6 # 76 – 103 Montería, Córdoba, Colombia
[2] Departamento de Física y Electrónica, Universidad de Córdoba, Carrera 6 # 76 – 103 Montería, Córdoba, Colombia
[3] Grupo Avanzado de Materiales y Sistemas Complejos (GAMASCO), Universidad de Córdoba, Colombia.

E-mail: ocjm19@gmail.com

Abstract. The activity and selectivity of a catalyst (Au/γ–Al_2O_3) and the support (γ–Al_2O_3 cubic) in the oxidation reaction of cyclohexene in the liquid phase to obtain (epoxycyclohexane, 2-cyclohexen-1-ona, 2-cyclohexen-1-ol) were evaluated. The catalyst was prepared by the precipitation-deposition method from aqueous solutions of $HAuCl_4$, citric acid (2% m/v) and polyvinyl alcohol. For the adequacy of the support, starting from boehmite calcined at 773 K, γ–Al_2O_3 phase was obtained. The reaction products and the evaluation of the same were analyzed by gas chromatograph (GC), showing relevant results of the Au/γ–Al_2O_3 system with high conversion levels, mostly to products of 2-cyclohexen-1-one, 2-cyclohexen-1-ol with 31.17% and 25.89% selectivity respectively, and to a lesser proportion to epoxycyclohexane with 13.33% selectivity. Regarding to the system formed by the support, the results reported were 13.31% for 2-cyclohexen-1-ona, 4.28% for 2-cyclohexen-1-ol and 3-23% epoxycyclohexane.

1. Introduction

Reactions with supported metals are quite important in the production of many organic compounds [1]. Oxidation of cyclohexane is one of the methods for the synthesis of chemical intermediates which are of great value in the agrochemical and pharmaceutical industries [2].

Obtaining intermediates involved in the oxidation reaction of cyclohexene, lead to very long reaction times and unwanted products which means lost in industrial processes. One way to increase the performance of chemical processes in industrial scale reactions, involves the use of catalysts.

Many heterogeneous catalysts have been evaluated on the oxidation reaction of cyclohexene, within the noble metals supported on porous solids and metalloporphyrins transition metal in homogeneous or supported medium [3]. Gold is very versatile as a catalyst to generate interest products in the chemical and pharmaceutical industry, with high efficiency and less toxic precursors promotes certain reactions at very low metal loadings and small particle sizes [4]. Catalysts of supported gold have been studied for a wide range of oxidation reactions including oxidation of CO, epoxidation of propylene, direct synthesis of hydrogen peroxide from oxygen and hydrogen, oxidation of cyclohexane to (Ketone/Alcohol) KA oil, among others [5].

2. Experimental

The catalyst was prepared by the method of "precipitation-deposition from aqueous solutions of $HAuCl_4$, citric acid (2% mass/volume) and polyvinyl alcohol. Starting from bohemite to obtain the support, it was calcined at 773 K in order to form the γ–Al_2O_3 phase with surface area BET of 2.56 $x10^{-1}$ m^2Kg^{-1}. The amount of deposited gold was determined by atomic absorption, the supported system was calcined at 573 K in air. The surface area and distribution of pore sizes of the catalyst were obtained from adsorption-desorption isotherms of nitrogen at 77 K and transmission electron microscopy (TEM) for dispersion of the metal particles.

Retention times of each component of the reaction are determined using hexane as an internal standard and dichloromethane as solvent at concentration ranges of 0.1, 0.2, 0.3, 0.4 and 0.5 N. In a reactor 3.6 $x10^{-4}$ L cyclohexene 0.35 N was added; 1.4 $x10^{-4}$ L of Hexane 0.1 N; 2 $x10^{-4}$ Kg of iodosylbenzene (PhIO) as an oxidizing agent; 5 $x10^{-5}$ Kg of catalyst and 7.5 $x10^{-3}$ L of dichloromethane. The reaction system was subjected to constant stirring at 298 K and 1 atm for 1.4 $x10^{4}$ s, taking samples of the reaction every 2.4 $x10^{3}$ s and immediately injecting 2 $x10^{-6}$ L in the gas chromatograph for monitoring the reaction. The reaction products were quantified by gas chromatography (GC) using a DB-5 column, of semi polar character and hexane as the internal standard, under the following conditions: initial temperature of 323 K and final temperature of 433 K, at a heating rate 1.67 $x10^{-1}$ K/s and held at 433 K for 3 $x10^{2}$ s with nitrogen as a carrier gas, and flow 2.1 $x10^{-4}$ L/s.

The area of the chromatographic peaks was analyzed by the method of internal standard (hexane), the formation epoxycyclohexane, 2-cyclohexen-1-ona, and 2- cyclehexen-1-ol was determined, formed from the oxidation reaction and the selectivity of the catalyst.

3. Results and discussion

The phase identification was carried out by X-ray diffraction using the X'pert HighScore 2.1 program, and the Match! program (Crystal impact). The XPowder program was also used to facilitate the study of diffraction X-ray obtained by the powder method, considering the data base PDF2-2004.

Figure 1 show X-ray diffractograms for uncalcined and calcined boehmite (773 K) for 4 hours. For the calcined sample, characteristic peaks of the gamma (cubic) phase at $2\theta = 48.90 - 49.03$ and $2\theta = 64.50 - 64.68$ positions can be observed. The analysis of the XRD presents characteristic patterns of boehmite phase (AlO (OH) 21-1307 (ICDD) and γ–Al_2O_3 phase(50-741 (ICDD) [6]. The results of chemical analysis by atomic adsorption for γ–Al_2O_3 and Au/γ–Al_2O_3 systems were 0,000 and 0,010 % mass of gold respectively.

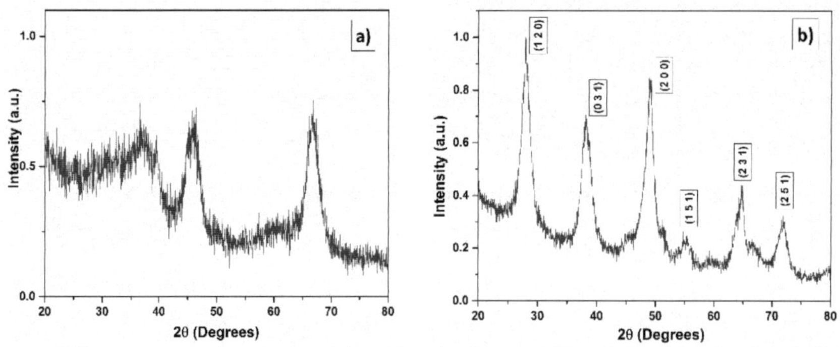

Figure 1. X-ray diffractograms for a) uncalcined boehmite and b) calcined boehmite (773 K)

Figure 2 corresponds to the adsorption-desorption isotherms of type IV nitrogen for the support γ–Al_2O_3, and the catalyst Au/γ–Al_2O_3 synthesized. Additionally, are shown: the surface areas, calculated by the BET model and pore size distributions by the BJH model. The support has an area of 2.56 $x10^{-1}$ m^2Kg^{-1} and average pore size of 58 Å according to Figure 4a. The system comprised of the Au/γ–Al_2O_3 has a 2.33 $x10^{-1}$ m^2Kg^{-1} area and an average pore size of 56 Å.

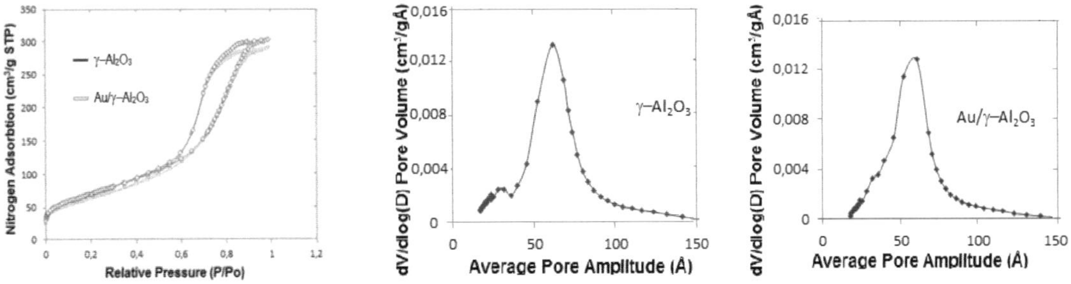

Figure 2. Isotherms of nitrogen adsorption of the catalyst and support γ–Al_2O_3 and pore size distribution for the γ–Al_2O_3 support and Au/γ–Al_2O_3.

Additionally, transmission electron microscopy (TEM) image of the catalyst Au/γ–Al_2O_3 catalyst (Figure 3), evidence a particle size of less than 20 nm.

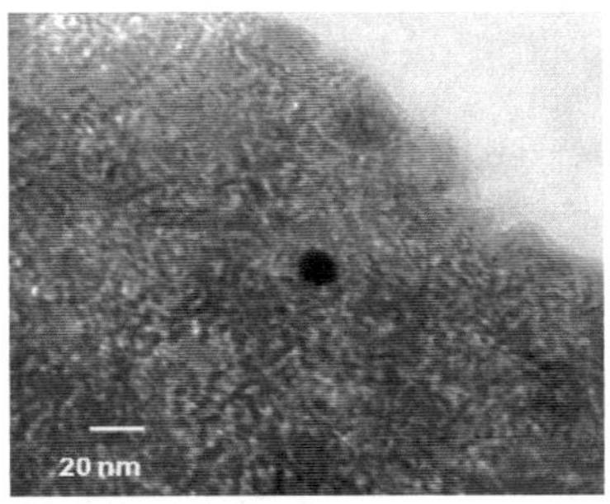

Figure 3.TEM image of the Au/γ–Al_2O_3 catalyst.

The products obtained in the evaluation of the catalytic activity in the oxidation reaction of cyclohexene with respect to support γ–Al_2O_3 were: epoxycyclohexane, 2-cyclohexene-1-ol and 2-cyclohexene-1-ona which are summarized in Table 1. All reactions were performed with 3.6 $x10^{-4}$ L of cyclohexane at a concentration of 0.35 N; 1.4 $x10^{-4}$ L of Hexane 0.1N ; 2 $x10^{-4}$ Kg of iodosylbenzene (PhIO) as an oxidizing agent; 5 $x10^{-5}$ Kg of catalyst and 7.5 $x10^{-3}$ L of dichloromethane as solvent.

Table 1. Catalytic evaluation on cyclohexane oxidation for catalysts

System	Time (min)	Conversion	Selectivity (%) Epoxidation		Via Allylic
			Cy-Epoxy	*Cy-1-ol*	*Cy--1-ona*
γ–Al_2O_3	240	20.83	3.23	4.28	13.31
Au/γ–Al_2O_3	240	70.40	13.33	25.89	31.17

Having account conversion of cyclohexane and the selectivity of the products to the oxidation reaction, the results exhibit a higher catalytic performance for Au/γ–Al_2O_3 respecting to γ–Al_2O_3, presenting higher levels of conversion to products 2-cyclohexen -1-ona, 2-cyclohexen-1-ol and a lower proportion of epoxycyclohexane.

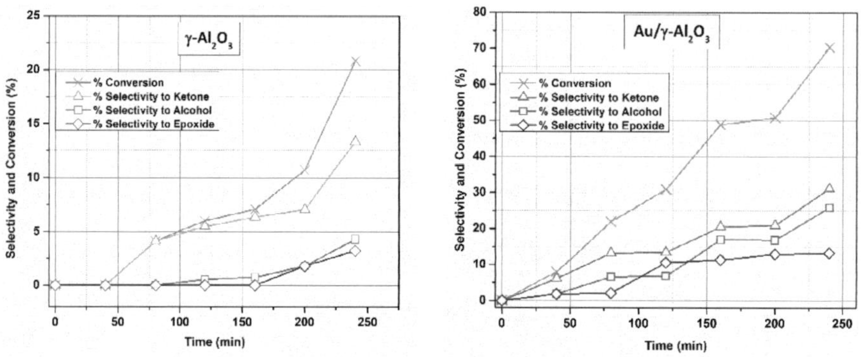

Figure 4. Results of the catalytic activity for the γ–Al_2O_3 and Au/γ–Al_2O_3 systems.

Figure 4 shows the results of the catalytic activity for γ–Al_2O_3 and Au/γ–Al_2O_3 systems. At $4.8x10^3$ s (80 min) of reaction, γ–Al_2O_3 promotes allylic oxidation, showing selectivity for 2-cyclohexen-1-one. At $7.8x10^3$ s (120 min), the catalytic activity is observed towards allylic oxidation and epoxidation. At $1.44x10^4$ s (240 min) of reaction, the conversion is 20.83%, with percentages of 3.23%, 4.28% and 13.31% for epoxycyclohexane, 2-cyclohexen-1-ol and 2-cyclohexen-1-ona, respectively. For Au/γ–Al_2O_3, a tendency to the formation of allylic oxidation during the reaction is observed, with some epoxidation products. At end of reaction, a conversion of 70.40% is obtained with selectivity rates of 13.33%, 25.89% and 31.17% for epoxycyclohexane, 2-cyclohexen-1-ol and 2-cyclohexen-1-ona respectively.

During the process, a marked competitiveness between the two reactions is observed, this behavior can be attributed to the nature of the catalyst used. The oxidation of allylic carbon occurs by influence of the increased acidity of the catalyst [7], where the alumina acts as a support of Au, presenting an appreciable catalytic activity due to its surface acidity and resulting intermediates reactive molecules adsorbed on metal support interfaces [1].

4. Conclusions

In this work, products obtained in the oxidation reaction of cyclohexene due to catalytic activity and selectivity of the γ–Al_2O_3 and Au/γ–Al_2O_3 were identified and quantified.
The catalytic activity of the Au/γ–Al_2O_3 system is increased compared to γ–Al_2O_3 support, associated with the presence of gold particles, possibly related to surface acidity and removing of OH groups.

References

[1]. Caballero C, Valencia J, Barrera M and Gil A 2010, *Powder Technol.* **203(2)** 412
[2]. Zhen-Yu C, et al 2011, *Adv. Che. Eng Sci.* **1** 15.
[3]. Karimipour G, Rezaei M, and Ashouri D 2013, *J. Mex. Chem. Soc* **57(4)** 276.
[4]. Kouachi K, Lafaye G, Especel C, Cherifi O and Marécot P 2008, *J. Mol. Catal. A. Chem* **280** 52.
[5]. Xu Li-Xin, et al 2008, *Catal. Commun* **9(5)** 816.
[6]. Paglia G et al 2004, *Chem. Mater* **16** 220.
[7]. Radman et al 2015, *Orient. J. Chem* **31(4)** 2277.

Ferroelectric response of $Na_{0.9}Li_{0.1}NbO_3$ at room temperature

V Isaza-Zapata[1], A Arias[1], C Maya[1], W Martínez[1], A Agudelo[1], B Álvarez[1], A Gómez[2] and J L Izquierdo[1,3*].

[1] Institución Universitaria Pascual Bravo, Facultad de Ingeniería, Laboratorio de Materialografía, Grupo GIIEN, A.A. 6564, Calle 73 No. 73A-226, Medellín, Colombia

[2] Universidad Nacional de Colombia, Facultad de Minas, Escuela de Materiales y Minerales, A.A. 568, Medellín, Colombia

[3] Universidad Nacional de Colombia, Facultad de Ciencias Básicas, Grupo de Materiales Cerámicos y Vítreos, A.A. 568, Medellín, Colombia

**Corresponding author: jorge.izquierdo@pascualbravo.edu.co, jlizquierdon@unal.edu.co*

Abstract. Polycrystalline samples of the dielectric, ferroelectric $Na_{0.9}Li_{0.1}NbO_3$ were synthesized by means of solid-state reaction using different ramps of temperature. High purity precursors Nb_2O_5, Na_2CO_3 and Li_2CO_3 were stoichiometricaly mixed and heated at 1373 K for 6 hours in air atmosphere. The $Na_{0.9}Li_{0.1}NbO_3$ samples were characterized by means of diffraction of X-rays using a Panalytical X'Pert Pro diffractometer. The obtained diffractograms were refined by Rietveld procedure using the software Maud. The results show that $Na_{0.9}Li_{0.1}NbO_3$ crystallizes in an orthorhombic structure (space group *Pbma*). The refined diffraction patterns also suggest a possible replacement of the Na ions by Li ions in the parent, ferroelectric $NaNbO_3$ compound. For electrical characterization of the samples, electrical polarization curves were measured using a Radiant Technologies RT66A test system for ferroelectric materials. The polarization versus applied electrical dependence shows a hysteretic behavior verifying the ferroelectric character of the material. A maxima polarization of 130 $\mu C/m^2$ is obtained at room temperature.

Keywords: Niobates; Solid state reaction; Perovskites; Polarization; Ferroelectricity.

1. Introduction

Up to 2003, the most popular piezoelectric material was the PZT ($Pb(Ti, Zr)O_3$). Nevertheless, this material was restricted in the European Union because the lead content, which is a dangerous substance for the health and environment [1,2,3]. The discovery of very promising piezoelectric properties in perovskite-type solid solutions of $(K, Na)(Nb, Ta)O_3$ has enabled the development of technologically important lead-free piezoelectric materials and stimulated renewed scientific interest in other members of these solutions, particularly $NaNbO_3$ (NNO) [4]. A direct mixture of NNO and $LiNbO_3$ generates the interesting material $Na_{1-x}Li_xNbO_3$ (NLN). NLN is considered as an advanced ferroelectric ceramic for high frequency applications [5,6]. Intensive studies on alkaline niobates-based new dielectric materials have been carried out by many groups around the world in order to obtain suitable materials for high performance applications. Although NNO is ferroelectric only around 163 K, the NLN system has been found to be ferroelectric at room temperature for doping levels $x \geq 0.02$ [7,8]. On the other hand, recent studies have shown that lithium niobate ($LiNbO_3$) is a very important material for optical storage, electric wave guides, amplitude modulators, second harmonic generators and acoustic wave transducers applications. The structure and electrical properties of the crystalline $LiNbO_3$ are different from those of NNO. At room temperature, $LiNbO_3$ is ferroelectric and NNO anti-ferroelectric with orthorhombic distorted perovskite-type structure. Furthermore, whereas NNO undergo successive phase transitions from low temperatures through high temperatures (10–1500 K), $LiNbO_3$ remains in the same state until 1523 K [9]. The aim of the present work is basically the study of the effect of doping with Li ions on the electrical properties of the $NaNbO_3$ compound in bulk at room temperature.

2. Experimental

Polycrystalline samples of NLN were synthesized by standard solid-state reaction. High purity reactants of Nb_2O_5, Na_2CO_3 and Li_2CO_3 were stoichiometricaly mixed and thoroughly grinded in an agate mortar. The resulting powder was slowly heated up to 1323 K for 2 h at a rate 3° C/min in air atmosphere, and then cooled to room temperature. The total reaction involved in the procedure can be written as [9]

$$\frac{(1-x)}{2}Na_2CO_3 + \frac{1}{2}Nb_2O_5 + \frac{x}{2}Li_2CO_3 \rightarrow Na_{1-x}Li_xNbO_3 + \frac{1}{2}CO_2 \qquad (1)$$

The powders were pressed in form of pellets (6 mm in diameter and thickness ~1 mm) using pressures as high as 100 bar. The resulting pellets were slowly heated in air (~5° C/min) up to 1373 K and then sintered there for 6 h. When the process was completed, the samples were cool down to room temperature in the furnace. The crystal structure of the pellets was analyzed by X-ray diffraction (XRD) using Cu Kα radiation (λ=1.5406 Å) in a standard θ–2θ configuration. Crystallographic parameters were obtained through Rietveld refinement using the software Maud. For electrical characterization of the samples, electrical polarization curves were measured using a Radiant polarimeter for ferroelectric materials. A standard bipolar wave with period of 10 ms was applied to cylindrical pellets of 6 mm in diameter and thicknesses of about 1mm. These pellets were placed between copper electrodes simulating a parallel plate capacitor.

3. Analysis and results

XRD patterns of the polycrystalline $Na_{1-x}Li_xNbO_3$ (x=0 and 0.1) samples were recorded at room temperature (Figure 1(a)). The diffraction patterns were simulated by the Rietveld method with the Maud software (crystallographic card 0018684) [10]. The theoretical fit of the XRD data for the doped sample is also presented in Figure 1(b). The results of this procedure indicated that the sintered ceramic samples have an orthorhombic crystal structure (space group *Pbma*) with lattice parameters a = 5.5655 Å, b = 5.5039 Å and c =15.5131 Å. These values resulted to be smaller than those reported for the undoped sample [11]. This result suggests a successful replacement of Na ions by Li ions in the structure of the parent compound [12]. Certainly, the XRD peaks of the doped sample (x=0.1) are shifted to the right side as compared to those of the NNO compound (see inset of Figure 1(a)). The shift of the peaks to higher angles suggest that the interplane distances decrease. Indeed, the presence of smaller Li ions generates relaxation of the oxygen ions surrounding it, which, in turn, leads to tilting of the oxygen octahedra. Although the XRD patterns show only an additional peaks stemming from impurities, it is known XRD, being a bulk analytical technique, would hardly detect small molar concentrations of impurities. As a fact of matter, the niobates are pretty difficult to prepare. During the synthesis, the phase $NaNb_3O_8$ may form easily. Certainly, the peak detected in the XRD spectrum seems to be originated by $NaNb_3O_8$ (<4%) [13]. Nevertheless, this secondary phase is not ferroelectric at room temperature.

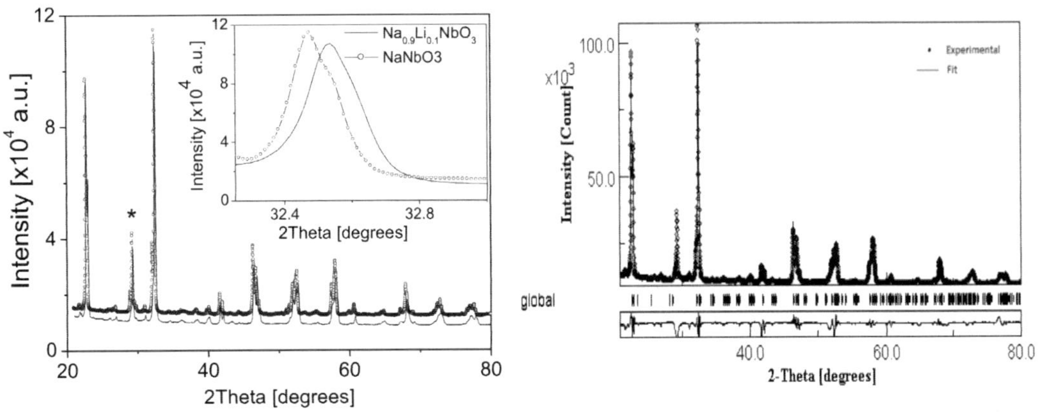

(a) (b)

Figure 1. (a) X-ray powder diffraction patterns of polycrystalline $Na_{1-x}Li_xNbO_3$ (x=0 and 0.1) samples synthesized by standard solid state reaction. Inset: XRD diffractograms plotted on an enlarged scale around the main peaks. Here, the symbol (*) represents the secondary phase denoted by $NaNb_3O_8$ (b) Rietveld refinament for the sample with x=0.1.

Measurements of polarization as a function of the applied electric field (Figure 2) on undoped and Li-doped samples evidence a hysteretic feature, which is the characteristic of the ferroelectric materials. The ferroelectric behavior of the materials are clearly observed at different voltages. Although the values of the electrical polarization of $Na_{0.9}Li_{0.1}NbO_3$ are certainly too small for technical applications, the results suggest that enhancement of the electrical properties of NNO is possible upon Li doping.

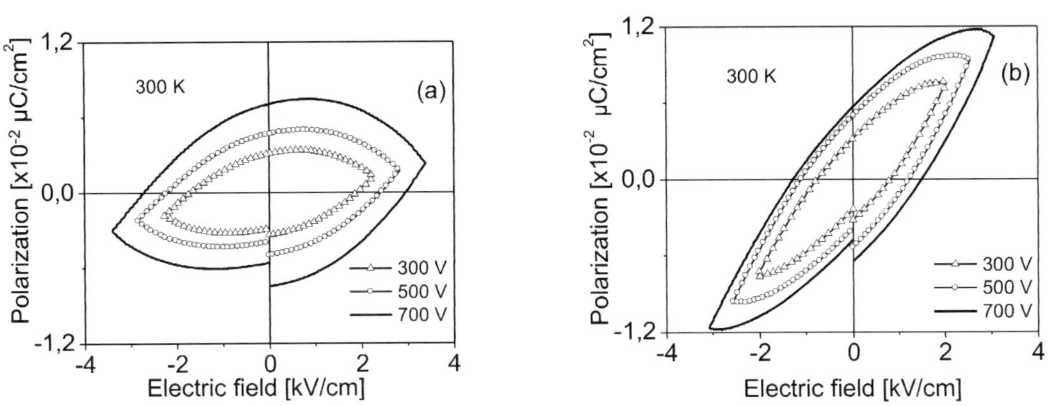

Figure 2. (a) Polarization as a function of the electric field for $NaNbO_3$ and (b) $Na_{0.9}Li_{0.1}NbO_3$ at various voltages.

Conclusions

Polycrystalline $Na_{0.9}Li_{0.1}NbO_3$ samples were successfully synthesized by standard solid-state reaction. This synthesis method allows to this material with orthorhombic structure. The replacement of the Na ions by Li ions was verified by XRD measurements. Low concentration of impurities was also confirmed by this technique. Electric polarization measurements at room temperature showed ferroelectric behavior of the compound. The results are encouraging in the sense that $Na_{0.9}Li_{0.1}NbO_3$ could represent a new class of ferroelectric materials working at room temperature.

Acknowledgments

The authors want to acknowledge the collaboration of Dr. O. Morán at the Universidad Nacional de Colombia, Medellín Campus for the polarization measurements.

References

[1] EU-DIRECTIVE-2002/95/EC, 2003, Official Journal of the European Union, **46** 19.

[2] EU-DIRECTIVE-2002/96/EC, 2003, Official Journal of the European Union, **46** 24.

[3] Guodong Shi, Junhan Wang, Hengli Wang, Zhanjun Wu, Huaping Wu, 2017, Ceram. Int. **43** 7222.

[4] Yasuyoshi S, Hisaaki T, Toshihiko T, Tatsuhiko N, Kazumasa T, Takahiko H, Toshiatsu N, Masaya N, 2004, Nature, **432** 84.

[5] Zeyfang R, Henson R, Maier W, 1977, J. Appl. Phys.**48**, 3014.

[6] Sadel A, Von Der M uhll R, Sadel A, Ravez J, Hagenmuller P, 1982, Solid State Commun. **41** 345.

[7] Von Der M uhll R, Sadel A, Ravez J, Hagenmuller P, 1979, Solid State Commun. **60** 15.

[8] Zhiqing L, Kun Z, Xiaoming L, 2011, Ferroelectrics-Physical Effects (intechopen), Ch. 28.

[9] Chaker C, Gharbi W, Abdelmoula N, Simon A, Khemakhem H, Maglione M, 2011, J. Phys. Chem. Solids, **72** 1140.

[10] Sakowski-Cowley A, Lukaszewicz K, Megaw H, 1969, Acta Cryst, **25** 851.

[11] Gusakova L, Ishchuk V, Kisel N, Kuzenko D, Spiridonov N, 2010, Fun. Mater. **17** 528.

[12] Pan H, Zhu G, Chao X, 2011, Mater. Chem. Phys. **126** 183.

[13] Kravchenko O, Gadzhiev G, Omarov Z, Reznichenko L, Abdullaev K, Razumovskaya O, Shilkina L, Komarov V, Verbenko I, 2011, Inorg. Mater, **47** 919.

IOP Conf. Series: Journal of Physics: Conf. Series **850** (2017) 012010 doi:10.1088/1742-6596/850/1/012010

Field calibration of volcanic surveillance cameras.

C A Ospina[1,2], A Narvaez[1,3] and I D Corchuelo[1,4]

[1] Colombian Geological Service, Volcanological and Seismological Observatory of Popayán, 190002, Colombia.

[2] cospina@sgc.gov.co, [3] anarvaez@sgc.gov.co, [4] icorchuelo@sgc.gov.co

Abstract. In volcanic surveillance, cameras are largely used allowing amazing images of volcanic eruptions as well as beautiful views of these grand Earth constructions. The Colombian Geological Service through the Volcanological and Seismological Observatory of Popayán (OVSPo) have 10 surveillance cameras looking at three volcanoes present in Provinces of Cauca, Huila and Tolima. However, these cameras were not calibrated previously, which has limited the analysis and exploitation of the information up to now. The development of this work take into account that the calibration process should not change camera parameters like orientation and position and what's more, we consider the access difficulties to reach and stay at the camera stations (volcanic environment). A calibration methodology was developed and applied on three (3) cameras on field, achieving to improve the analysis and exploitation of information within images of volcanic surveillance cameras.

1. Introduction

Nowadays volcanic surveillance through images is largely used, based on satellites, overflights, in-situ pictures and images from surveillance cameras [1], including spectral ranges as IR, UV [2] and visible. Each one of these techniques have benefit, drawbacks and different complexity to get relevant information about volcanic activity. Concerning images achieved by surveillance cameras in visible spectra, the Colombian Geological Service (SGC) through the Volcanological and Seismological Observatories (from Manizales-OVSM, Pasto-OVSP and Popayán-OVSPo) has surveillance cameras installed on field on strategic places to register surface volcanic activity, and so with radio link is transmitted one image per minute. Once they are at office, analysis is established according with all techniques of volcanic surveillance (seismology, chemistry, deformation...), so a better comprehension of volcanic activity is achieved.

Until now, analysis methodology used include activities like visual identification and anomaly measurements as ash emissions, glows, pyroclastic flows, volcanic eruptions... Measurements at the OVSPo had been achieved based on known real points and its correspondence on the image [3], so when using cross multiplication or an analogical scale in a transparent film, they had estimates of the real dimensions in meters. Other way is the use of pixel measurements of phenomena. Despite of its useful these methods possess unknown accuracy because perspective, depth, distortions problems among others. In the OVSPo these problems have been addressed by some works [4][5] and approached by teams in other countries [6].

Camera calibration is a process that allows finding the relation between a 3D real point and a 2D image point; it can lead to distortion reduction because lenses and getting knowledge of camera installation parameters like orientation. Calibration methods with patterns [7] had been used on volcanic surveillance cameras like on Etna volcano Italy [8] to estimate eruption column height, in which known patterns are located in front of the camera for calibration. Although currently there is a broad variety of methods to facilitate this process, they depend on scene conditions sight by camera,

camera characteristics or the possibility to controllably and slightly move the camera [9], what is not easy on field cameras.

In the OVSPo, the cameras did not have calibration at installation moment, so measurements from its images have allowed seeing relative changes in time but not getting a coherent analysis with real phenomena measures. In this document, a field camera calibration method is presented and the results on three (3) of the 10 surveillance cameras in the OVSPo what has led to better volcanic phenomena analysis through image analysis.

2. Volcanic surveillance cameras at OVSPo

From 2008, when the OVSPo installed its first surveillance camera, it has increased the number of surveillance cameras up today we count with five looking at the Nevado del Huila volcano, four at the Puracé volcano and one at the Sotará volcano. Those cameras are installed on strategic places but geographically hard access, requiring up three days by foot to arrive the station, under severe weather conditions (figure 1). We took into account those conditions to create and apply the Calibration method.

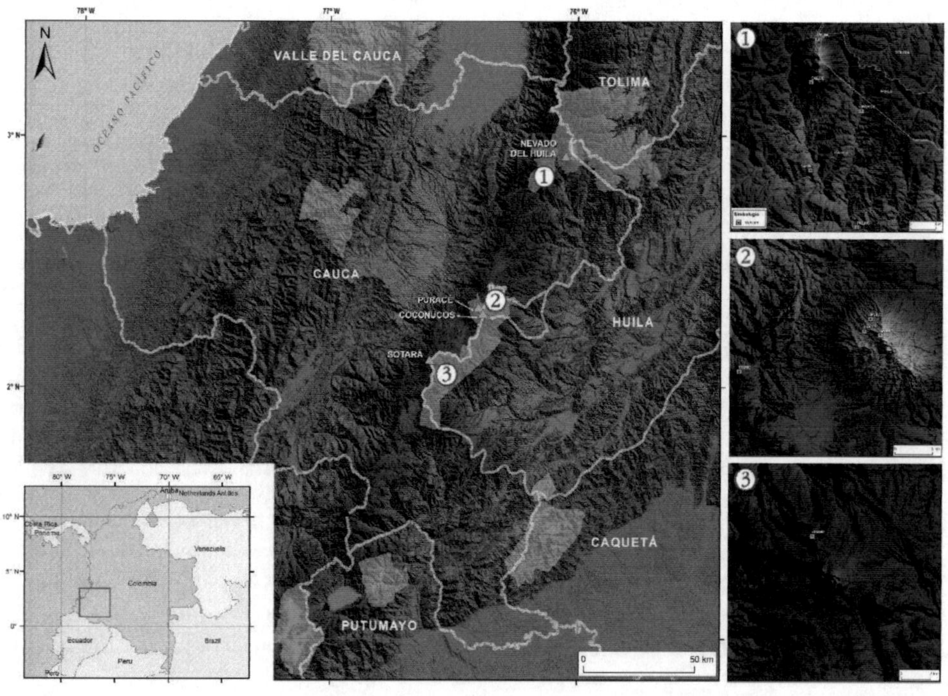

Figure 1. Location map of 1) Nevado del Huila, 2) Puracé and 3) Sotará volcanoes and its Net of Surveillance cameras at SGC (orange squares).

3. Traditional measurements methods of volcanic surface phenomena

On volcanic surveillance, is too much important to know the quiet times of the volcano and then make a comparison of time behaviour, to identify anomalies associated with volcanic changes. Regarding to volcanic surveillance cameras, observations and measurements on images, the situation is the same, it is important to know the volcanic behaviour through parameters like normal column height and its colouring.

Based on one image, is possible to develop measurements in pixels or meters if there is a calibrated camera. Pixel measurements are useful to identify anomalies through time, but off course is not enough to understand the real volcanic behaviour.

Other way to measure adapted at some volcanoes, is the use of a vertical analogical scale in a transparent film (figure 2). This scale fits in the same observation point at which a person is located to see the volcano through the transparent film and then make measurements based on the scale and the phenomenon behind. We saved a record data at the end.

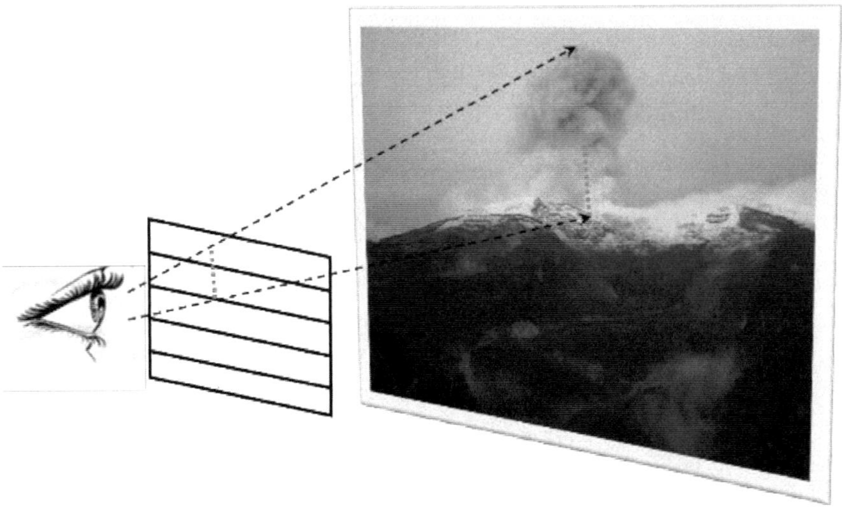

Figure 2. Measurements through analogical scale in a transparent film (left side) of volcanic phenomena (right side).

Last case is the use of known coordinate points that are visible on images, to get an approximate scale on the image. So using cross multiplication, they got an estimation of phenomenon real dimensions.

In all cases relative measurements are recorded, what allows identifying changes in volcanic activity; however, only real measurements serve to some institutions in charge of control and regulation aviation (Aerocivil en Colombia). Furthermore, all cases involve several problems that leads to unknown errors, problems like depth and parallax.

4. Field calibration of volcanic surveillance cameras

Camera calibration intend to find the existing relation between object coordinates and image coordinates. Therefore, it is possible to position an object or develop measurements on the image and calculate the real position and real measurements, based on image data.

One of the methods to develop real measures through images is based on pin-hole camera model and the knowledge of orientation, vertical and horizontal field of view (FOV), camera matrix and distortion parameters [10]. Method used on some volcanic surveillance cameras, like at Etna volcano [8] and selected for project too.

There are several approaches to achieve the camera matrix and distortion parameters, but here was used the chessboard pattern [7]. Furthermore, to get orientation and FOV was invented a pattern tool ("T") to be carried to camera stations on field. Next figure shows the pattern.

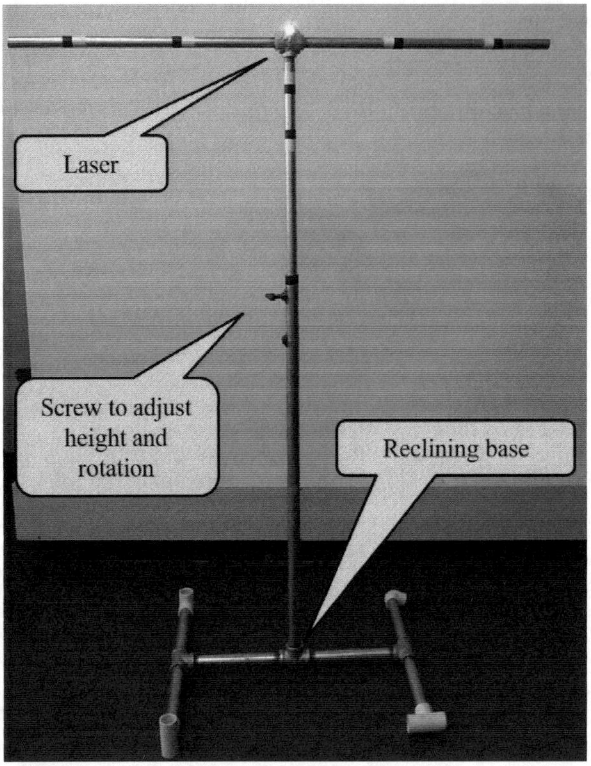

Figure 3. "T" pattern instrument. Yellow brands at T pattern are at known distance each other.

We also took into account the following facts to create the pattern:

- Field cameras should not be moved or even touched while calibration, because the results must serve to analysis older images.
- Field cameras could be installed on masts up to 4 meter height and "T" pattern should be in front of cameras.
- Pattern must be rigid enough to stand firm despite windy, damp and rainy weather.
- Technically, the pattern should be adjustable in such a way it forms a parallel plane to image plane. Therefore, the pattern need mobile parts to modify height, tilt and rotation.

That is why we thought of using galvanized pipe nipple forming a T what should be adjusted to be parallel to the image plane, helped by a laser in the pattern middle, perpendicular to T pattern, laser which will point to camera lenses (parallel to camera sight line) to validate that pattern is parallel to the image plane.

The proposed field camera calibration proceeding that considers next steps:

In office, before going field, the staff should identify and draw the image centre, because this point and the laser must match while calibrating.

Once at field:

1. Request to the office personnel, some snapshots of chessboard pattern in front of camera.
2. Locate T pattern in front of camera and make to fit the T pattern centre (laser) with image centre, before identified.
3. Turn on the laser and point it to camera lenses. Ensure second step is satisfied too.
4. Then, request to the office personnel some snapshots of T pattern in front of camera.
5. Followed, should be fastened a thread to the T pattern middle. Then it is stretched to reach the camera lenses or to set parallel to camera line sight.
6. On this thread, camera elevation and azimuth are measured.
7. In addition, thread distance (Dh) is needed for next processes.

Once the stuff returns to the office, may find the camera matrix and distortion parameters. We, compute the FOV through equations (1) for vertical axis and (2) for horizontal axis, represented in figure 4.

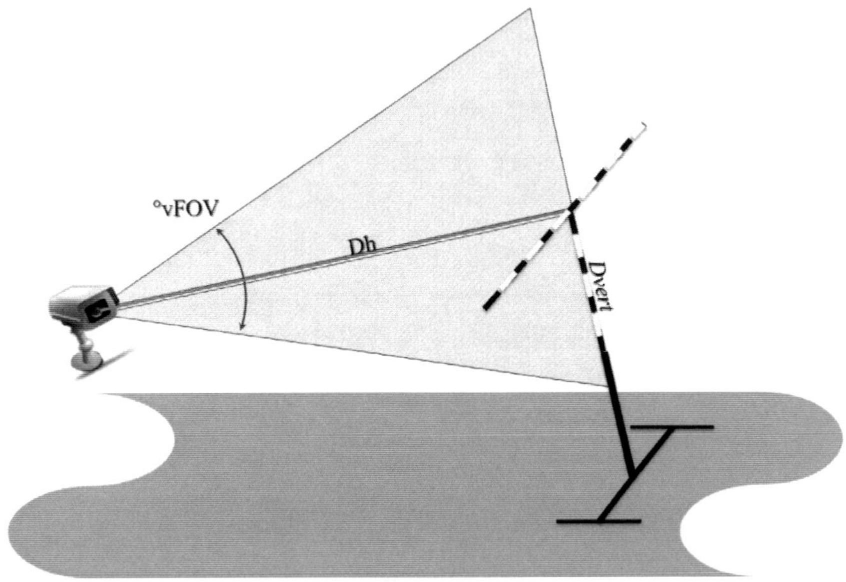

Figure 4. Process scheme of field calibration of surveillance cameras. Yellow marks at T pattern are at known distance each other. Red line indicates the pointing of laser.

$$vFOV = 2 * \tan^{-1}\left(\frac{D_{vert}}{2 * Dh}\right) \quad (1)$$

$$hFOV = 2 * \tan^{-1}\left(\frac{D_{horiz}}{2 * Dh}\right) \quad (2)$$

Now is possible to get the meters to pixels rate for each axis for a fixed depth *d*, what depends on image resolution, as equation (3) shows:

$$MetersPixRate = \frac{d * \tan(^{FOV}/_{2})}{^{resolution}/_{2}} \ \frac{[\mathrm{m}]}{[\mathrm{pix}]} \quad (3)$$

Therefore, it is possible to know the dimensions of an object on the image plane at a depth *d* from camera, multiplying the size of the object, in pixels, by the meters pixels rate for each axis, once the image is undistorted and rotated.

IOP Conf. Series: Journal of Physics: Conf. Series **850** (2017) 012010 doi:10.1088/1742-6596/850/1/012010

5. Results and conclusions

We tested the calibration process in three of the ten surveillance cameras of OVSPo. Next figure show some images recorded during calibration process of these three surveillance cameras and the results are in table 1.

Figure 5. Field calibration camera at three of ten volcanic surveillance cameras. a) Lavas Rojas (LARWC), b) Caloto (CALWC) and c) Mina (MNAWC).

Table 1. Calibration results in the three tested surveillance cameras tested at OVSPo.

	Caloto (CALWC)	**Mina (MNAWC)**	**Lavas Rojas (LARWC)**
Azimuth	35	138	100
Elevation	20	16	20
Rotation	8.5±0.5	-1.7±0.5	0±0.5
hFOV	60±2.1	34±1.3	72±2
vFOV	37.7±6.6	26.6±5	57.8±2.3

Orientation parameters: azimuth, elevation and rotation, in addition to horizontal and vertical field of view were found at three of the ten OVSPo volcanic surveillance cameras, reaching errors up to 18% of the measure in vFOV of Mina camera, what's is not so good yet. These measures represent the beginning on calibration methods at these cameras and will be used to record better phenomena measures than before, however there are still some challenges to be faced:

- Improve accuracy on all parameters.
- The installation of each camera changes its conditions and the T pattern might not be helpful in all of them.
- Some cameras are on masts of up to 4 meters of height and could oscillate with wind and

therefore orientation parameters could change, so measures on these images could have unpredictable errors.

- The zoom or focus at each camera could change and if it is modified, the calibration process should be applied again. Because of that, cameras with motorized lenses controlled remotely could not be exploited fully.
- The same situation happens when station maintenance require moving the camera.

Therefore, a versatile method should be the next step to take advantage of all situations and so achieve a better support to understand volcanic activity state. To achieve this, one possibility is using digital compass that could be incorporated to the camera installation and then know all the time the position. In addition, the physical location of a fixed pattern in front of each camera that would allow knowing the changes in orientation, zoom and focus in a certain time.

Bibliography

[1] B. Andò and E. Pecora, "An advanced video-based system for monitoring active volcanoes," *Comput. Geosci.*, vol. 32, no. 1, pp. 85–91, 2006.

[2] U. Platt, P. Lübcke, J. Kuhn, N. Bobrowski, F. Prata, M. Burton, and C. Kern, "Quantitative imaging of volcanic plumes - Results, needs, and future trends," *J. Volcanol. Geotherm. Res.*, vol. 300, pp. 7–21, 2014.

[3] "Boletín semestral de actividad de los volcanes Nevado del Huila , Puracé y Sotará segundo semestre de 2008," Popayán, 2008.

[4] C. A. Ospina and L. Pencue, "Flow measurement in open channels based in digital image processing to debris flow study," *Comput. Model. Objects Represent. Images Fundam. Methods Appl. III*, pp. 49–52, 2012.

[5] L. C. (INGEOMINAS) Manzo Oscar, Santacoloma Cristian, "Análisis de cambios superficiales asociados a la actividad eruptiva en el volcán nevado del Huila entre 2007 y 2010," in *XIV Congreso Latinoamericano de Geología*, 2011.

[6] S. A. Valade, A. J. L. Harris, and M. Cerminara, "Plume Ascent Tracker: Interactive Matlab software for analysis of ascending plumes in image data," *Comput. Geosci.*, vol. 66, pp. 132–144, 2014.

[7] G. Bradski and A. Kaehler, *Learning OpenCV, Computer Vision with OpenCV Library*, O'Reilly. O'Reilly Media, Inc., 2008.

[8] S. Scollo, M. Prestifilippo, E. Pecora, S. Corradini, L. Merucci, G. Spata, and M. Coltelli, "Eruption column height estimation of the 2011-2013 Etna lava fountains," *Ann. Geophys.*, vol. 57, no. 2, 2014.

[9] J. I. Gonzalez, "Estudio experimental de métodos de calibración y autocalibración de cámaras," LAS PALMAS DE GRAN CANARIA, 2003.

[10] A. de la. Escalera Hueso, *Visión por computador: fundamentos y métodos*, Pearson Ed. Prentice Hall, 2001.

Acknowledgments

To God, my family and my wife Jazmín. To the staff of the Volcanological and Seismological Observatory of Popayan, who supported at all times the development of this work.

Global characterization of nematic liquid crystal display Sony LCX038ARA with applied electric field in the modulation amplitude-coupled regime

C J Cuevas Cely, C H Acevedo Cáceres and Y Torres Moreno
GOTS, Optics & Signal Processing Group, Physics School, Sciences Faculty, Universidad Industrial de Santander, 680002 Bucaramanga, Colombia.

E-mail: cuevasjennifer07@gmail.com

Abstract. In this work, we showed experimental and theoretical results of the characterization of a spatial light modulator based in a nematic liquid crystal display Sony model LCX038ARA with electric field excitation. The parameters: effective molecular twist and equivalent retardation are determinated using the retarder-rotor technique in the amplitude-coupled regime.
Keywords. SLM, LCD, Amplitude-coupled Regime of Modulation, Retarder-rotor Technique, Nematic phase SLM.

1. Introduction

One of the main applications of Liquid Crystal Display (LCD) devices is in Spatial Light Modulators (SLMs) [1-2]. Most SLMs are in nematic phase and are known as Twisted Nematic Liquid Crystal Displays (TN-LCDs). TN-LCDs produce amplitude-coupled modulation depending on the applied voltage [3].

To characterize the complete behavior of a TN-LCD is necessary to know the magnitude and sign of the manufacture parameters of the LCD with applied electric field between their pixelated electrodes, like the effective molecular twist and the equivalent retardation [4]. In this work, we show the experimental calibration of a transmission LCD Sony LCX038ARA for the previous cited parameters. Both parameters are found using the retarder-rotor technique. This technique is a combination of the Soutar-Lu method for a null voltage applied to LCD and their Stokes parameters [5-6]. In addition, we analyze the experimental results obtained previously for the molecular twist angle ($\emptyset$), the angle of the molecular axis (Ψ_D), and birefringence (β) when a null voltage is applied [7].

2. Methodology

To calibration of a TN-LCD under applied voltage, we found the total twist angle ($\emptyset$), the angle of molecular axis at the entrance face (Ψ_D) and the maximum birefringence (β), without applied voltage used the Soutar-Lu method [7]. The Soutar-Lu model uses the Jones matrix analysis when the TN-LCD is sandwiched between two polarizers with the purpose of find the parameters above mentioned it using the transmitted intensity curves for the crossed polarizers and parallel polarizers.

After finding these parameters, we proceed to characterize the TN-LCD by the retarder-rotor technique. This method considers that the TN-LCD behaves as a polarization system formed by a retarder followed by a rotor, through an applied voltage. The retardation effect is added on an arbitrary phase state [8]. Both $\emptyset_{eq}$ and δ_{eq} in equation (1) take different values depending on the magnitude of the voltage applied to TN-LCD. External voltage on the TN-LCD is accomplished by transforming the gray levels to 0-255 scale potential differences through the photoelectric effect, for the case of an 8 bits converter. At each gray level *g* corresponds a voltage V and the relationship between the two values is usually linear [5]. $\emptyset_{eq}$ and δ_{eq} are estimated using circularly polarized light. The LCD in terms of retardation and rotor is written as [6,8]:

$$\boldsymbol{M}_{TN-LCD} = exp(-i\beta)\boldsymbol{R}(\emptyset_{eq})\boldsymbol{WP}\left(2\delta_{eq},\frac{\emptyset+\emptyset_{eq}}{2}\right), \quad (1)$$

where β and $\emptyset$ are the birefringence and molecular twist respectively, when the TN-LCD has a zero voltage applied. While δ_{eq} is the equivalent retardation and $\emptyset_{eq}$ is the effective twist for each potential difference. Then, the electromagnetic wave at the output of the TN-LCD has a electric field amplitude E_{out} that can be written by the following equation [6,8]:

$$E_{out} = exp(-i\beta)\boldsymbol{R}(\emptyset_{eq})\boldsymbol{WP}\left(2\delta_{eq},\frac{\emptyset+\emptyset_{eq}}{2}\right)E_{in}. \quad (2)$$

with E_{in}, the incoming polarized electric field, typically circular polarized; exp $(-i\beta)$ is the inherent retardation of the birefringent material, $\boldsymbol{R}(\emptyset_{eq})$ is a rotor operator or twist and $\boldsymbol{WP}(2\delta_{eq}, (\emptyset + \emptyset_{eq}) / 2)$ is the retardation operator that describes the state of polarization due to the external voltage applied. Cartesian components of the electric field E_{out} lead to find mathematical expressions for Stokes parameters of the light transmitted through the TN-LCD given by [6,8]:

$$S_0 = 1, \quad (3)$$
$$S_1 = \sin(\emptyset - \emptyset_{eq})\, sin2\delta_{eq}, \quad (4)$$
$$S_2 = -cos(\emptyset - \emptyset_{eq})\, sin2\delta_{eq}, \quad (5)$$
$$S_3 = cos2\delta_{eq}. \quad (6)$$

According to equation (6) we can infer that it is possible to find the equivalent retardation (δ_{eq}) using the S_3 parameter. While, the S_1/S_2 ratio could lead to determine the equivalent twist ($\emptyset_{eq}$), both when an external voltage is applied to the TN-LCD [6,8]:

$$\delta_{eq} = \frac{1}{2}arccos(S_3) \quad (7)$$
$$\emptyset_{eq} = \emptyset + arctan\left(\frac{S_1}{S_2}\right) \quad (8)$$

3. Results

The table 1 shows the experimental results obtained for the total twist angle ($\emptyset$), the angle of molecular axis at the entrance face (Ψ_D) and maximum birefringence (β) of the TN-LCD with null applied voltage.

Table 1. Measured parameters for the TN-LCD without applied voltage.

Parameter	
Twist angle ($\emptyset$)	+93.8°
Angle of the molecular axis (Ψ_D)	-14.9°
Birefringence (β)	2.126 rad

After finding the initial parameters of the TN-LCD, we proceed to find the experimental values of $\emptyset_{eq}$ and δ_{eq} using the schematic setup shown in Fig. 1. The light source is a He-Ne laser (L) model R30990 from Newport™ operating at 632.8 [nm] with 5 [mW] of power and random polarization. In order to eliminate high spatial frequencies like noise from the emerging light beam, we have used the objective (O) and pinhole (PH) to spatial filter the laser beam. Subsequently the emerged beam is collimated using the lens L1. The choice of lens L1 was made taking into account the selected spatial filter and searching a full illumination of the TN-LCD active area. The LCD is sandwiched between a polarizer (P) and an analyzer (A). Finally, the output beam converges at the image focal plane of lens L2 with the purpose to obtain efficient intensity measurements of the output beam using the power-meter (PM). Setup includes two quarter wave plates (QWP or λ/4). The first quarter wave plate (QWP1) is placed in front of TN-LCD to generate the required circular polarization state. While the second one (QWP2) is placed after the TN-LCD for S_3 measurement parameter (see equation 6). Thus $\emptyset_{eq}$ and δ_{eq} both are experimentally obtained from the Stokes parameters following a method analogous to described in previous section, for a series of 26 images of gray levels ranging from g = 0 to g = 255 in steps of 10 displayed on the TN-LCD [9].

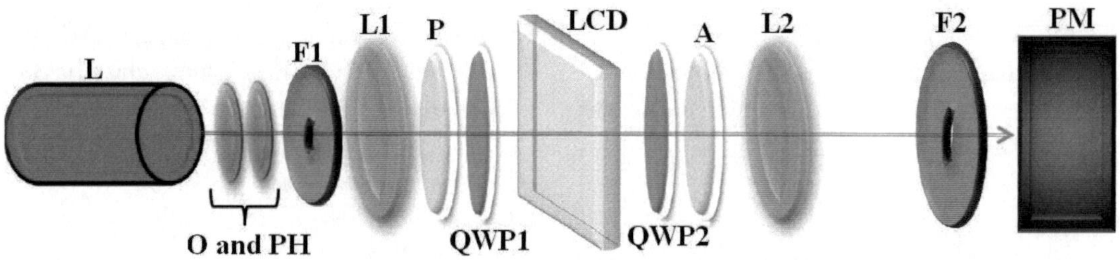

Figure 1. Schema of Experimental setup for measuring $\emptyset_{eq}$ and δ_{eq} using circularly polarized light to illuminate the TN-LCD.

The left side of Fig. 2 shows the experimental results for S_1, S_2 and S_3 and for each gray level g applied to the TN-LCD model LCX038ARA. From the experimental curve shown on the left of Fig.2 and equations (7) and (8) are found the values of $\emptyset_{eq}$ and δ_{eq}. The right side of same figure shown the experimental values of $\emptyset_{eq}$ and δ_{eq}. In this figure we can see that the experimental values for effective twist shows abrupt jumps for the first seven measured gray levels. This behavior can be explained by the corresponding uncertainty of $\emptyset_{eq}$, given by [10]:

$$\sigma_{\emptyset_{eq}} = \frac{1}{S_1^2+S_2^2}\sqrt{S_2^2\sigma_1^2 + S_1^2\sigma_2^2}, \tag{9}$$

where σ_1 and σ_2 are estimates uncertainties for S_1 and S_2 Stokes parameters. According to equation (9), the uncertainty of $\emptyset_{eq}$ will grow unlimited when S_1 and S_2 tend to zero or are very small, as seen for the first gray levels of experimental plot shown on the left side of the Fig. 2. For this reason, the accurate determination of $\emptyset_{eq}$ for the first gray levels requires another method. The value

IOP Conf. Series: Journal of Physics: Conf. Series **850** (2017) 012011 doi:10.1088/1742-6596/850/1/012011

of $\emptyset_{eq}$ for the first gray levels can be found if the TN-LCD is illuminated with linear polarized light in the direction of the horizontal axis of the crystal reference system itself, and consequently the corresponding normalized intensity is written as [6]:

$$T = \begin{pmatrix} E_x \\ E_y \end{pmatrix} \begin{pmatrix} E_x^* & E_y^* \end{pmatrix} = cos^2\delta_{eq}cos^2(\xi + \emptyset_{eq}) + sin^2\delta_{eq}cos^2(\xi - \emptyset_{eq}). \quad (10)$$

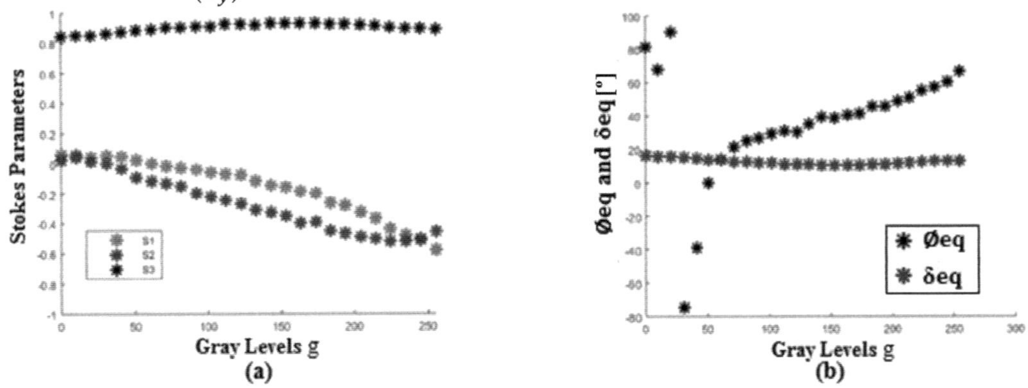

Figure 2. Experimental results for the Stokes parameters S_1, S_2 and S_3 with circular polarized light throughout of TN-LCD in terms of the gray level g deployed (left side). Experimental results for the values $\emptyset_{eq}$ and δ_{eq} (right side).

where ξ is the angle between the transmission axis of the analyzer and the molecular director axis in front of the TN-LCD. From equation (10) is possible to determine the value of $\emptyset_{eq}$ and δ_{eq} by a nonlinear fit of the experimental values of $T(\xi)$. The left side of Fig. 3 shows the experimental results for transmission TN-LCD at zero gray level together with the plot obtained for nonlinear fitting. From this curve we infer the value of effective twist and equivalent delay, given by $\emptyset_{eq} = 12.000°$ and $\delta_{eq} = 16.036°$, respectively. Following an analog procedure, we obtained the values of $\emptyset_{eq}$ and δ_{eq} for other six remaining gray levels.

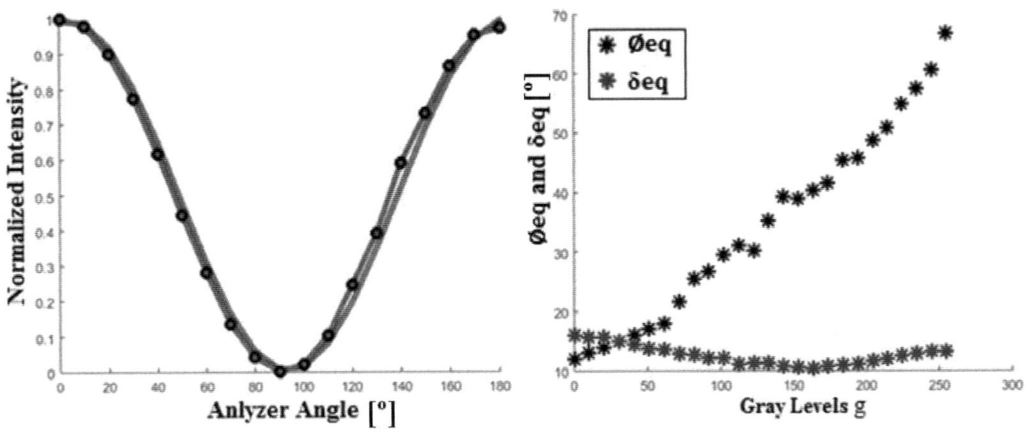

Figure 3. Experimental results (blue points) and nonlinear fitting (red curve) to find $\emptyset_{eq}$ and δ_{eq} at the zero gray level (left side). Adjusted experimental results for δ_{eq} and $\emptyset_{eq}$ versus gray level applied to TN-LCD (right side).

Finally, the right side of Fig. 3 shows the experimental data for equivalent retardation (blue plot) and effective twist (black plot) in function of gray level displayed onto TN-LCD (external applied voltage). From these data, we can see that the TN-LCD has a higher performance as rotor than as retarder, i.e. our TN-LCD Sony LCX038ARA has a tendency global to rotate the polarization state of the incident beam that it includes an additional phase value on the incident light.

4. Conclusions

The retarder-rotor technique combined with Stokes parameters could be used for characterize any TN-LCD when an external potential difference is applied. Parameters characterizing the equivalent system (δ_{eq} and ϕ_{eq}) for each gray level were experimentally measured. Both δ_{eq} as ϕ_{eq} are estimated from measurements of the Stokes parameters using incoming circularly polarized light to the TN-LCD. Comparing the obtained results of the transmitted light of the TN-LCD when a linear polarized wave and a circular polarized electromagnetic wave is used, we have found that the values for ϕ_{eq} and δ_{eq} are very similar. However, it is worth noting that the scheme of lighting the TN-LCD to find the effective twist and equivalent retardation when an external potential difference is applied is very tedious regarding the method of Stokes parameters.

5. Acknowledgments

Research partially supported by Universidad Industrial de Santander with the projects 5708 and 1788. Thanks to Colciencias for financial support to project 110256934957: “Dispositivos ópticos para sistemas de alta dimensionalidad QKD (Quantum Key Distribution) basados en el momento angular orbital de la luz”. C.H. Acevedo Cáceres acknowledges the support and sponsorship from Call 647-2015: National Doctoral program of Colciencias.

6. References

[1] T. Inaba, R. Eguchi, and H. Tsuda, “Focusing grating coupler for integration of a liquid-crystal spatial light modulator onto a Si photonic circuit,” In *Frontiers in Optics*. Optical Society of America, paper JTh2A-148, 2016, October.

[2] J. R. Lindle, A. T. Watnik, and V. A. Cassella, “Efficient multibeam large-angle nonmechanical laser beam steering from computer-generated holograms rendered on a liquid crystal spatial light modulator,” *Applied Optics*, *55* (16), pp.4336-4341, 2016.

[3] J. M. Vilardy, M. S. Millán and E. Pérez-Cabré, “Static and dynamic amplitude modulation of light in a twisted nematic liquid crystal display,” *Proc. SPIE - Int. Soc. Opt. Eng.*, Vol. 8785, November, 10pp., 2013.

[4] H. F. Varó, “*Estudio y optimización del almacenamiento de información en una memoria holográfica*,” PhD Tesis, Universidad de Alicante, 2008.

[5] C. Soutar and K. Lu, “Determination of the physical properties of an arbitrary twisted-nematic liquid crystal cell,” *Opt. Eng.*, Vol. 33, No. 8, p. 2704, 1994.

[6] V. Durán Bosch, "Optimización del funcionamiento de un modulador espacial de luz de cristal líquido mediante el modelo retardador-rotor. Aplicaciones en óptica adaptativa.", PhD Tesis, Universidad de Valencia, 191 pp., 2007.

[7] H. K. Liu and T. H. Chao, "Liquid-Crystal Television Spatial Light Modulators," *Applied Optics*, Vol. 28, No.22, pp. 4772-4780, 1989.

[8] V. Duran et Al., “Equivalent retarder-rotator approach to on-state twisted nematic liquid crystal displays,” *Journal of Applied Physics,* Vol. 99, #113101, 6pp., 2006.

[9] R. H. Chen, “Liquid crystal displays: fundamental physics and technology,” John Wiley & Sons, 2011.

[10] P. R. Bevington and D. K. Robinson, “Data Reduction and Error Analysis for the Physical Sciences,” 2nd Ed., New York: Mc Graw-Hill, 1992.

Image encryption using the Gyrator transform and random phase masks generated by using chaos

Juan M. Vilardy, Carlos J. Jimenez, and Ronal Perez

Grupo GIFES. Faculty of engineering. Universidad de La Guajira, Riohacha (La Guajira), Colombia

E-mail: jmvilardy@uniguajira.edu.co

Abstract. The Gyrator transform (GT), chaotic random phase masks (CRPMs) and a random permutation of the Jigsaw transform (JT) are utilized to design an images encryption-decryption system. The encryption-decryption system is based on the double random phase encoding (DRPE) in the Gyrator domain (GD), this technique uses two random phase masks (RPMs) to encode the image to encrypt (original image) into a random noise. The RPMs are generated by using chaos, these masks are CRPMs. The parameters of the chaotic function have the control of the generation of the CRPMs. We apply a random permutation to the resulting image of the DRPE technique, with the purpose of obtaining an encrypted image with a higher randomness. In order to successfully retrieve the original image (without errors or noise-free) at the output of the decryption system is necessary to have all the proper keys, which are: the rotation angles of the GTs, the parameters of the chaotic function utilized to generate the two CRPMs and the random permutation of the JT. We check and analyze the validity of the image encryption and decryption systems by means of computing simulations.

1. Introduction

The image encryption is different from traditional cryptology because the spatial, frequency and redundancy features of the image to encrypt are analysed and processed by the image encryption system [1]. The double random phase encoding (DRPE) is a successful method for optical image encryption [1–3], the DRPE uses two random phase masks (RPMs) with the purpose of encoding the image to encrypt (original image) into a stationary white noise pattern (encrypted image). One of the main drawbacks of the image encryption systems based on the initial DRPE is that the security of the system is vulnerable to attacks, this weakness is due to the linear property of the DRPE scheme [1–3]. The DRPE has been further extended from the Fourier domain to the Fresnel domain [4–6], the fractional Fourier domain [7–12], the Gyrator domain (GD) [13–16] and other domains [17–20], with the purpose of adding more keys and increasing the security of the DRPE system.

The Gyrator transform (GT) is a mathematical tool for analysis and processing of two-dimensional signals [21]. The GT has been used in optics [22], signal processing [13] and image encryption [13–16].

In this paper, we propose a nonlinear image encryption-decryption system based on the DRPE, the GT, the chaotic random phase masks (CRPMs) and the Jigsaw transform (JT), in order to overcome the security vulnerabilities of the initial DRPE proposed in Ref. [2]. We use the GT to improve the security of the original DRPE by adding a new key for the encryption system (the rotation angle of the GT). It has been demonstrated that the nonlinearities introduced in the initial DRPE leading to an improvement in security [3, 5, 11, 14]. The generation of the RPMs by using a chaotic map allows to obtain the chaotic RPMs or CRPMs with a higher security. We use the Logistic map in order to generate the random codes of the CRPMs [23]. The values of the generated random codes are highly sensitive to the values of the parameters that control the iterations of the Logistic map [23]. This high sensitivity on the parameters of the Logistic map is used to obtain keys with a high security. The generation of the CRPMs by using the Logistic map is a nonlinear process. Finally, we use the JT in order to convert the encrypted image more random and to make the encryption/decryption system becomes nonlinear.

The outline of the paper is as follows: the definition and properties of the GT are introduced in section 2. The process to generate the random code of the CRPM by using a chaotic map is presented in section 3. The JT is introduced in section 4. In section 5, the nonlinear encryption and decryption systems are described. Simulation results for the encryption and decryption systems are provided in section 6. Finally, conclusions are summarized in section 7.

2. The Gyrator transform

The Gyrator transform (GT) at parameter α, called further as a rotation angle, of a two-dimensional function $f_i(x_i, y_i)$ is defined as

$$f_o(x_o, y_o) = \mathscr{G}^{\alpha}\{f_i(x_i, y_i)\} = \int_{-\infty}^{+\infty}\int_{-\infty}^{+\infty} f_i(x_i, y_i) K_\alpha(x_i, x_o, y_i, y_o) \mathrm{d}x_i \mathrm{d}y_i, \quad (1)$$

with

$$K_\alpha(x_i, x_o, y_i, y_o) = \frac{1}{|\sin(\alpha)|} \exp\left[i2\pi \frac{(x_o y_o + x_i y_i)\cos(\alpha) - (x_i y_o + x_o y_i)}{\sin(\alpha)}\right], \quad -\pi < \alpha \leq \pi, \quad (2)$$

where K_α is the kernel function of GT. The GT is mathematically defined as a linear canonical integral transform which produces the rotation in position–spatial frequency planes (x, q_y) and (y, q_x) of phase space [21]. For $\alpha = 0$, it corresponds to the identity transform. For $\alpha = \pm\pi/2$ it reduces to the direct/inverse Fourier transform with rotation of the coordinates at $\pi/2$. For $\alpha = \pm\pi$ the reverse transform described by the kernel $\delta(x_i \pm x_o, y_i \pm y_o)$ is obtained. The GT is periodic and additive with respect to rotation angle: the period is 2π and $\mathscr{G}^{\alpha}\mathscr{G}^{\beta} = \mathscr{G}^{\alpha+\beta}$, respectively. The inverse GT corresponds to the GT at angle $-\alpha$.

The main theorems such as scaling, shift, modulation, etc. for the GT have been formulated in ref. [21]. The optical GT can be implemented using the optoelectronic setup described in ref. [22]. The digital GT will be computed using the fast algorithm of discrete GT based on convolution operation [24] for the nonlinear image encryption-decryption system that it will be described in section 5.

3. Random phase masks (RPMs) generated by using chaos

The RPMs for the encryption and decryption systems are generated by using a chaotic map. We use the Logistic map in order to generate the random codes of the CRPMs. The Logistic map is a one-dimensional chaotic map defined as

$$x_{n+1} = \lambda x_n (1 - x_n), \quad (3)$$

where $\lambda \in (0, 4)$, $x_n \in [0, 1]$ and $n = 0, 1, 2, 3, \ldots$ [23]. The Logistic map given by the Eq. (3) is a recursive relation that it uses one initial value for x_n, which is x_0, and one value for λ. When the values $x_0 \in [0, 1]$ and $\lambda \in [3.6, 4)$, the sequence x_n is chaotic [23].

(a)

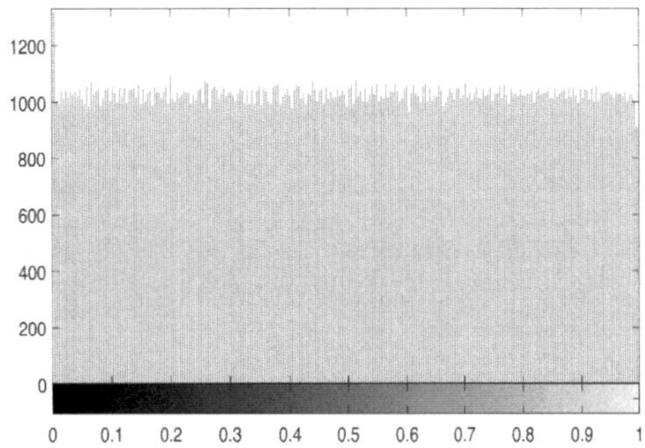

(b)

Figure 1. (a) Random code generated by using the Logistic map, and (b) Histogram of the random code generated in (a).

The random codes of the CRPMs are random real-valued images of $M \times N$ pixels. In order to generate a random code of $M \times N$ pixels, we set the values of $x_0 \in [0, 1]$ and $\lambda \in [3.6, 4)$, and the Eq. (3) is iterated n times. The value of x_{n+1} will be the first value of the random code. Then, the Logistic map given by the Eq. (3) is iterated $MN - 1$ times. The generated values between x_{n+1} and x_{n+MN} represent the values of the random code when these generated values are reshaped into a two-dimensional array of $M \times N$ pixels. All the generated pixels have their values in the interval [0, 1]. Therefore, we need three parameters (n, x_0 and λ) with the purpose of generating a random code of $M \times N$ pixels by using the Logistic map.

A generated random code of 512×512 pixels for the parameters $n = 121$, $x_0 = 0.712745$ and $\lambda = 3.89945$ is shown in figure 1(a). The histogram of the generated random code is presented in figure 1(b). The pixel values of the generated random code by using the Logistic map follow the values of an uniform distribution, this type of random distribution is commonly used in the optical encryption systems based on the DRPE [1,2].

The generation of the random code by using the Logistic map or other chaotic map is more convenient to distribute or to store the RPMs keys of an encryption system, because it is not necessary to distribute or to store all the $M \times N$ pixels of the random code. We just need to provide three parameters in order to generate a random code of $M \times N$ pixels by using the Logistic map.

4. The Jigsaw transform

The Jigsaw transform (JT) is defined as a juxtaposition of different sections of an image [25]. A simple two-dimensional case is shown in figure 2(a). The effect of this transform on an image (figure 2(b)) is shown in figure 2(c).

In this case the image was broken up into 64 subsections of 8×8 pixels size, which were repositioned relative to each other according to some permutation. The permutation used is random. The JT is unitary, energy being conserved through the transform and it also has an inverse. In the case shown in figure 2(c), there are 64! possible permutations for the JT. Each particular JT is denoted by some index e.g. $J_b\{\}$ and its inverse is denoted by $J_{-b}\{\}$.

5. Nonlinear image encryption and decryption systems

In this section, we describe all the processes about the encryption and decryption systems. The image to encrypt (original image) $f(x, y)$ is a real-valued distribution and their pixel values are

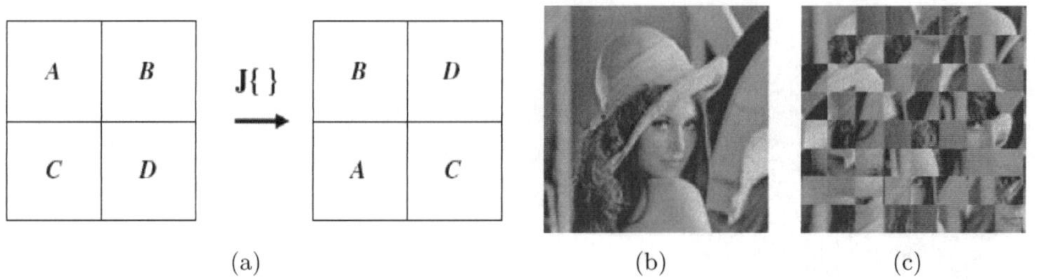

Figure 2. (a) A simple case of the JT. (b) A 256 × 256 pixels image in grayscale (levels ranging from 0 to 255) used as an input. (c) Input image after the first 16 × 16 JTs.

in the the interval [0, 1]. The CRPMs $r(x, y)$ and $h(u, v)$ are given by

$$r(x, y) = \exp\{i2\pi s(x, y)\}, \quad h(u, v) = \exp\{i2\pi n(u, v)\}, \tag{4}$$

where (x, y) and (u, v) represent the coordinates for the spatial domain and the GD determined by the result of the GT at parameter α, respectively, $s(x, y)$ and $n(u, v)$ are random codes generated by using the Logistic map and the generation process described in section 3. All the pixel values of $s(x, y)$ and $n(u, v)$ are in the interval [0, 1]. To generate the random code $s(x, y)$, we use three parameters $(n^{(1)}, x_0^{(1)}, \lambda^{(1)})$ and for the random code $n(u, v)$, we utilize $(n^{(2)}, x_0^{(2)}, \lambda^{(2)})$.

First, we describe all the steps of the encryption system. The original image $f(x, y)$ is encoded in phase $f_{ph}(x, y) = \exp\{i2\pi f(x, y)\}$ [9, 14]. The image $f_{ph}(x, y)$ is bonded to the CRPM $r(x, y)$ and then, the GT at parameter α is performed

$$g_\alpha(u, v) = \mathscr{G}^\alpha\{r(x, y)f_{ph}(x, y)\}. \tag{5}$$

The complex-valued distribution $g_\alpha(u, v)$ is multiplied by the CRPM $h(u, v)$ and this product is transformed using the GT at parameter β. Finally, we apply the JT with an index b to the result of the last GT, in order to obtain the complex-valued encrypted image

$$e(p, q) = J_b\{\mathscr{G}^\beta\{h(u, v)g_\alpha(u, v)\}\} = e_a(p, q)\exp\{i2\pi e_{ph}(p, q)\}, \tag{6}$$

where (p, q) are the coordinates for the GD given by the result of the GT at parameter $\alpha + \beta$, the index b of the JT denotes a specific random permutation of T elements, the real-valued images $e_a(p, q)$ and $e_{ph}(p, q)$ are the amplitude and phase distribution of the complex-valued encrypted image $e(p, q)$. The CRPMs $r(x, y)$ and $h(u, v)$ are used to spread the information content of the original image $f_{ph}(x, y)$ onto the encrypted distribution $e(p, q)$ and to make this encrypted distribution a stationary white noise pattern [1, 2]. The JT provides a higher randomness to the encrypted image. The nine security keys of the encryption system are given by the two rotation angle α and β of the GTs, the six parameters $(n^{(1)}, x_0^{(1)}, \lambda^{(1)})$ and $(n^{(2)}, x_0^{(2)}, \lambda^{(2)})$ used to generate the random code $s(x, y)$ of the CRPM $r(x, y)$ and the random code $n(u, v)$ of the CRPM $h(u, v)$, respectively, and the random permutation of T elements used in the JT.

The decryption process is the same encryption process applied in the inverse sense on the complex-valued encrypted image $e(p, q)$ with the index $-b$ of the JT, the complex conjugate of the CRPMs $r(x, y)$ and $h(u, v)$ and the negatives rotation angles ($-\alpha$ and $-\beta$) for the GTs. Using the additivity property of the GT and the inverse of the JT, we obtain

$$d(x, y) = arg\{r^*(x, y)\mathscr{G}^{-\alpha}\{h^*(u, v)\mathscr{G}^{-\beta}\{J_{-b}\{e(p, q)\}\}\}\} = 2\pi f(x, y), \tag{7}$$

where the function $arg\{\cdot\}$ is used for extracting the argument of the resulting phase. The original image $f(x, y)$ can be correctly recovered without errors at the output of the decryption system whenever the values of the nine security keys used in the encryption and decryption systems are the same.

(a) $f(x,y)$ (b) $e_a(p,q)$ (c) $e_{ph}(p,q)$ (d) $d(x,y)$

(e) $d(x,y)$ (error α) (f) $d(x,y)$ (error $x_0^{(2)}$) (g) $d(x,y)$ (error index b)

Figure 3. (a) Original image to encrypt $f(x,y)$. Complex-valued encrypted image for the rotation angles $\alpha = 1.77\pi$ and $\beta = 0.41\pi$ of the GTs and the other seven security keys: (b) Amplitude image $e_a(p,q)$, and (c) Phase image $e_{ph}(p,q)$. (d) Decrypted image $d(x,y)$ using the right values of the nine security keys. Decrypted images for the following wrong security keys: (e) the rotation angle $\alpha = 1.770001\pi$, (f) the parameter $x_0^{(2)} = 0.245775001$, and (g) the random permutation used in the JT.

6. Computer simulations

The results of the computer simulations of the encryption and decryption systems following the steps described in section 5 are shown in figure 3. All image used in the encryption and decryption systems have 512×512 pixels in grayscale. The image to be encrypted (original image) is presented in figure 3(a). The random code $s(x,y)$ of the CRPM $r(x,y)$ generated by using the process described in section 3 with the parameters $n^{(1)} = 121$, $x_0^{(1)} = 0.712745$ and $\lambda^{(1)} = 3.89945$, it was depicted in figure 1(a). The random code $n(u,v)$ of the CRPM $h(u,v)$ was generated by using the Logistic map with the parameters $n^{(2)} = 577$, $x_0^{(2)} = 0.245775$ and $\lambda^{(2)} = 3.93217$. The random code $n(u,v)$ of the CRPM $h(u,v)$ has different values but similar appearance to the image presented in figure 1(a).

The amplitude and phase images of the complex-valued encrypted image for the rotation angles $\alpha = 1.77\pi$ and $\beta = 0.41\pi$ of the GTs, the previous generated CRPMs $r(x,y)$ and $h(u,v)$ and a random permutation of 1024 elements of the JT, are shown in figures 3(b) and 3(c), respectively. When the JT is applied over an image, this image is divided in blocks of 32×32 pixels and each block is repositioned by using the same random permutation of 1024 elements and the JT. The decrypted image $d(x,y)$ obtained from the complex-valued encrypted image $e(x,y)$ and the right values of the nine security keys is shown in figure 3(d).

To evaluate the quality of the decrypted images, we use the root mean square error (RMSE) between the decrypted images $d(x,y)$ and the original image $f(x,y)$ [3]

$$\text{RMSE} = \left(\frac{\sum_{x=1}^{M} \sum_{y=1}^{N} [f(x,y) - d(x,y)]^2}{\sum_{x=1}^{M} \sum_{y=1}^{N} [f(x,y)]^2} \right)^{\frac{1}{2}}. \tag{8}$$

IOP Conf. Series: Journal of Physics: Conf. Series **850** (2017) 012012 doi:10.1088/1742-6596/850/1/012012

The values of the RMSE metric for evaluating image quality are between 0 and 1; when the value of the RMSE is close or equal to 0, this metric indicates an excellent quality image for the retrieval of the decrypted image at the output of the decryption system whereas the values of the RMSE close or equal to 1 represent a worse quality image. The RMSE between the original image of figure 3(a) and the decrypted image of figure 3(d) is 4.4×10^{-32}. Therefore, the decrypted image of figure 3(d) is a perfect replica of the original image $f(x, y)$.

The random pattern images presented in figures 3(e), 3(f) and 3(g) correspond to the obtained decrypted images from the encrypted image $e(p, q)$ using a wrong value for the rotation angle α, the parameter $x_0^{(2)}$ or an incorrect random permutation used in the JT, respectively. The RMSEs between the original image of figure 3(a) and the decrypted images of figures 3(e), 3(f) and 3(g) are 0.81, 0.72 and 0.87, respectively. When the other six security keys (β, $n^{(1)}$, $x_0^{(1)}$, $\lambda^{(1)}$, $n^{(2)}$, and $\lambda^{(2)}$) used in the decryption process are not equal to the values used in the encryption process, the decrypted image is a noisy pattern very similar to the figures 3(e), 3(f) and 3(g). Therefore, all the nine security keys with their correct values are needed in the decryption system for the correct retrieval of the original image.

The key space analysis for the proposed nonlinear encryption system consists of every possible combination of the values of the nine security keys: the two rotation angles α and β of the GTs, the six parameters $(n^{(1)}, x_0^{(1)}, \lambda^{(1)})$ and $(n^{(2)}, x_0^{(2)}, \lambda^{(2)})$ used to generate the random code $s(x, y)$ of the CRPM $r(x, y)$ and the random code $n(u, v)$ of the CRPM $h(u, v)$, respectively, and the random permutation of 1024 elements used in the JT.

The GT has a period equal 2π with respect to the rotation angle α or β. From computational experiments it was found that the rotation angles α or β are sensitive to a variation of 1×10^{-7}. This sensitivity of the rotation angle of the GT was calculated using the values of the RMSE between the original image and the decrypted images for different rotation angles, including the right and wrong values of α and β. Therefore, the key space for the two rotation angles of the GTs is $4\pi^2 \times 10^{14}$.

The parameter n used in the generation of the random codes of the CRPMs is limited to the maximum value of 1×10^6, this value depends on the computing power used to generate the random code. The possibles values for $n^{(1)}$ and $n^{(2)}$ are 1×10^{12}. From computational experiments it was found that the parameters x_0 and λ are sensitive to a variation of 1×10^{-16} and 1×10^{-15}, respectively. The parameters $x_0 \in [0, 1]$ and $\lambda \in [3.6, 4)$. Therefore, the possibles values for $x_0^{(1)}$ and $x_0^{(2)}$ are 1×10^{32}, and the possibles values for $\lambda^{(1)}$ and $\lambda^{(2)}$ are 1.6×10^{29}

The random permutation of the JT has 1024 elements. Thus, there are 1024! possible permutations for the JT.

Finally, the total key space of the proposed nonlinear encryption process is of the order of $(4\pi^2 \times 10^{14})(1 \times 10^{12})(1 \times 10^{32})(1.6 \times 10^{29}) \cdot (1024!) = (6.3 \times 10^{88}) \cdot (1024!)$, this number represents a very larger key space. Hence, the brute force attacks are intractable just considering every possibility of the nine security keys.

7. Conclusions

A new nonlinear method for images encryption and decryption has been proposed involving the use of the double random phase encoding technique, the Gyrator transform (GT), the generation of random phase masks (RPMs) by using a chaotic map and the Jigsaw transform (JT). The nine security keys allow to obtain a very larger key space for the proposed security system. Therefore, the secret of the encrypted image is very-well protected against security attacks. The GT increases the security on the encrypted image due to the addition of a new key, given by the rotation angle. The generation of the chaotic RPMs or CRPMs by using the Logistic map is more suitable to distribute or to storage these security keys, because it is much easier to generate a random code of $M \times N$ pixels using a chaotic iterative map with three parameters instead of distributing or storing all the $M \times N$ pixels of the random code. The generation of the CRPMs is very sensitive with respect to the values of the three parameters used in the Logistic map. Apart from this, the random permutation used in the nonlinear JT increases much more the security for any cryptanalyst who tries to decrypt the digital image without authorization. The encrypted image is more protected due to the result of the JT and the all possible random permutations that can be used in the JT. Finally, the simulation results have shown that the encryption-decryption systems are more secure because the decrypted images are very sensitive to the change on the right values of the nine security keys.

Acknowledgments
This research has been funded by the Universidad de La Guajira.

References
[1] Millán M S and Pérez-Cabré E 2011 *Optical and Digital Image Processing: Fundamentals and Applications* ed G Cristóbal, P Schelkens and H Thienpont (Weinheim: Wiley) chapter 33 pp 739–67
[2] Réfrégier P and Javidi B 1995 *Opt. Lett.* **20** 767–9
[3] Vilardy J, Millán M S and Pérez-Cabré E 2013 *J. Opt.* **15** 025401
[4] Situ G and Zhang J 2004 *Opt. Lett.* **29** 1584–6
[5] Vilardy J, Millán M S and Pérez-Cabré E 2014 *Appl. Opt.* **53**, 1674–82
[6] Vilardy J, Millán M S and Pérez-Cabré E 2013 *Proc. of SPIE* **8785**, 87853J
[7] Unnikrishnan G, Joseph J and Singh K 2000 *Opt. Lett.* **25** 887–9
[8] Vilardy J, Torres C and Mattos L 2008 *AIP Conf. Proc.* **992** 1067–72
[9] Vilardy J, Torres C and Mattos L 2008 *Revista Colombiana de Física* **40** 143–6
[10] Vilardy J, Torres C and Mattos L 2011 *Revista Colombiana de Física* **43** 523–7
[11] Vilardy J, Torres Y, Millán M S and Pérez-Cabré E 2014 *J. Opt.* **16** 125405
[12] Vilardy J, Millán M S and Pérez-Cabré E 2014 *Opt. Pura y Apl.* **47** 35–41
[13] Rodrigo J A, Alieva T and Calvo M L 2007 *Opt. Commun.* **278** 279–84
[14] Vilardy J, Millán M S and Pérez-Cabré E 2016 Nonlinear image encryption using a fully phase nonzero-order joint transform correlator in the Gyrator domain *Opt. Laser Eng.* (*Preprint* http://dx.doi.org/10.1016/j.optlaseng.2016.02.013)
[15] Vilardy J, Torres C and Mattos L 2013 *Proc. of SPIE* **8785** 87851Q
[16] Vilardy J, Millán M S and Pérez-Cabré E 2015 *Proc. 20th Symposium on Signal Processing, Images and Computer Vision (STSIVA) (Bogotá)* vol 1 (Washington: IEEE Publishing) pp 1–6
[17] Morales Y, Díaz L and Torres C 2015 *J. Phys.: Conf. Series* **582** 012063
[18] Vilardy J, Torres C and Jimenez C 2013 *Proc. of SPIE* **8785** 87851R
[19] Vilardy J, Torres C, Useche J and Mattos L 2011 *J. Phys.: Conf. Series* **274** 012047
[20] Vilardy J, Torres C, Useche J and Mattos L 2011 *Revista Colombiana de Física* **43** 170–4
[21] Rodrigo J A, Alieva T and Calvo M L 2007 *Opt. Express* **15** 2190–203
[22] Rodrigo J A, Alieva T and Calvo M L 2007 *J. Opt. Soc. Am.* A **24** 3135–9
[23] Artur M 2002 *Chaos, Soliton & Fractals* **14** 1749–56
[24] Liu Z, Chen D, Ma J, Wei S, Zhang Y, Dai J and Liu S 2011 *Optik* **122** 864–7
[25] Hennelly B and Sheridan J 2003 *Opt. Lett.* **28** 269–71

IOP Conf. Series: Journal of Physics: Conf. Series **850** (2017) 012013 doi:10.1088/1742-6596/850/1/012013

$In_xGa_{1-x}As$ obtained from independent target via co-sputtering deposition

R Bernal-Correa[1,*], S. Torres-Jaramillo[2], C. Pulzara-Mora[2], J. Montes-Monsalve[2], S. Gallardo-Hernández[3], M. López –López[4], J. Cardona-Bedoya[5], A. Pulzara-Mora[2].

[1] Grupo DEMA, Facultad de Ciencias e Ingeniería, Universidad del Sinú, Montería Colombia.
[2] Laboratorio de Nanoestructuras Semiconductoras, Facultad de Ciencias Exactas y Naturales, Universidad Nacional de Colombia, Manizales, Colombia. A.A. 124.
[3] Departamento de Ingeniería Eléctrica, Centro de Investigación y de Estudios Avanzados del IPN, Avenida IPN No. 2508 07360 México D. F, México.
[4] Departamento de Física, Centro de Investigación y de Estudios Avanzados del I.P.N, Apartado Postal 14-740, 07000, México D.F, México.
[5] Grupo de Investigación en Materiales Semiconductores y Superiónicos, Departamento de Física, Universidad del Tolima, Ibagué, Colombia, AA 546.

robertobernal@unisinu.edu.co; aopulzaramo@unal.edu.co

Abstract. This paper is focused on the preparation of InGaAs thin films on GaAs substrates by r.f. magnetron sputtering technique, using the sputtering power as control means for the formation of different stoichiometries. Results of X-ray and Raman spectroscopy allowed corroborating the formation of $In_xGa_{1-x}As$ in different concentrations, identifying peaks associated with crystallographic planes (X-rays) and characteristic vibrational phonon modes (Raman). An analysis performed with the Secondary Ion Mass Spectroscopy (SIMS) and X-ray Photoelectron Spectroscopy (XPS) techniques, allowed discussing on the composition in each of the layers. Finally, an alternative in obtaining the ternary semiconductor with polycrystalline structure and preferential growth along the direction (111) was demonstrated and generated by a technique different from the epitaxial techniques, which are commonly used for the growth of III-V semiconductors.

1. Introduction

The use of non-epitaxial techniques for the preparation of ternary III-V semiconductors at laboratory scale has gained interest during the last decade due to a reduction of the costs that could be generated in the optoelectronic industry [1-5]. One of the most versatile and most commonly used techniques for the generation of materials in thin film form is magnetron sputtering, which makes it one of the most widely used technique for large scale materials production [6]. It is worthwhile mentioning that one of the most promising ternary III-V semiconductor is InGaAs, since it is possible to modify its optoelectronic properties by varying the concentration (x) of In, associated with the replacing of Indium atoms with

Gallium atoms in the GaAs matrix [7-9]. Accordingly, our efforts have been focused on the preparation of InGaAs thin films on GaAs substrates through magnetron sputtering technique, using sputtering power as a control means for different stoichiometries when depositing thin films. Evidence of said behavior is corroborated by structural, morphological and compositional characterization techniques.

2. Experimental

InGaAs thin films were deposited by magnetron sputtering r.f. on GaAs substrates (conventional cleaning) in an atmosphere of Argon at a pressure of 5 x 10^{-2} Torr for 60 minutes at a temperature of 580 ° C. During sputtering process, targets of Gallium Arsenide and Indium (two inches and one inch, respectively) were used. In the InGaAs growth (by co-sputtering technique), sputtering powers of 50 Watts for Gallium Arsenide and 4, 6, 8 Watts for the Indium target were used. The structural characterization of thin films was performed by means of the X-ray diffraction technique on a PANalytical Xpert Pro machine/instrument/system, equipped with a radiation source of Cu Kα (λ = 1.540562 Å) at a speed of 1 ° / min, in a range of 20 ° $\leq 2\theta \leq$ 40 °. Raman microscopy measurements were performed using a backscattering configuration, perpendicular to the sample, and using the 632.8 nm line of a He-Ne laser, in a Dilor XY Labram spectrometer, equipped with an Olympus BX40 microscope. The morphological characterization was made by Field Emission - Scanning Electron Microscopy (FE-SEM) at different magnifications, in a JEOL JSM740 1F instrument. Finally, a compositional study was conducted by using the XPS technique in an Alpha 10 Hemisphrical Analyser machine, equipped with an X-ray tube with double anode Mg Kα eV and Al Kα lines, which provides energies of 1253.6 eV and 1486.6 eV, respectively. SIMS analyzes were performed on a SIMS-IMS GF system, which is equipped with a primary beam energy of 5 keV with an approximate angle of 45 ° with respect to the sample. CsM+ secondary ions (M is the element of interest) were analyzed using a Double-focusing mass spectrometer equipped with a photomultiplier. The samples, in all cases, were scanned in an area of 100x 100 μm. The ions emitted from the central region of the sample were focused on the mass spectrometer.

3. Results

3.1. X-Ray

Figure 1 shows x ray spectra of InGaAs thin films deposited on GaAs substrates (100). These spectra show that the peaks (in 2θ) are shifted to smaller angles compared to the characteristic values of Gallium Arsenide, which is a good indicative of the formation of the ternary semiconductor. A preferential growth along the (111) direction is observed. Indium segregation in the semiconductor is discarded, reverse to the reported when it is deposited on other substrates [10]. The crystallite size of the layers was estimated using Scherrer formula using (111) plane (results are shown in the inset of Figure 1). Finally, a first approximation to the composition of ternary InGaAs was performed using Vegard's law [11]. Values of x = 7.73%, 17.0% and 19.50% for 4 W, 6 W and 8 W target powers were found.

3.2. Raman

In the first-order Raman spectra shown in Figure 2, TO and LO Raman modes of InAs and GaAs are observed. A deconvolution made in Raman spectra using Lorentzian functions allowed determining the position of each vibrational mode. In all spectra, a shift ($\Delta\omega$) in frequency (cm^{-1}) of the TO-InAs mode for the sample prepared at different Indium sputtering powers (4W, 6W and 8W) was observed, regarding to the TO mode (InAs bulk) located at 221 cm^{-1}. The shift ($\Delta\omega$) is a combined effect of stress due to parameter lattice mismatch between InGaAs layer and GaAs substrate, and allowing because the Ga atom are substituted by In atoms into GaAs matrix during growth. The Raman spectra show that the intensities ratio (I_{LO}/I_{TO}) of the LO and TO GaAs and InAs modes is larger than one, which is an indication that the samples have a high crystalline structure.

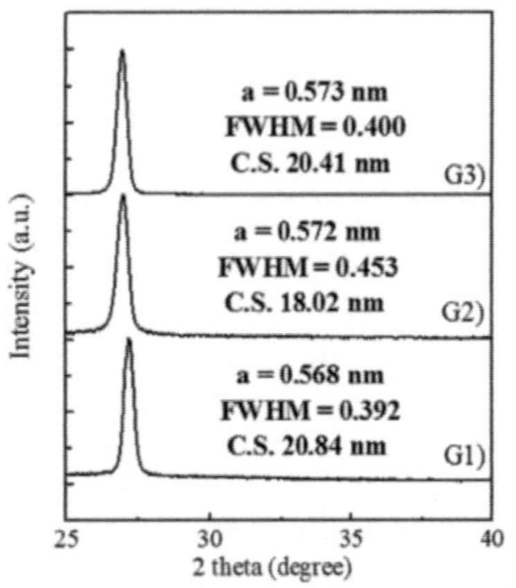

Figure 1. Diffractograms of InGaAs thin films varying the sputtering power of the Indium target at 4, 6, and 8 W.

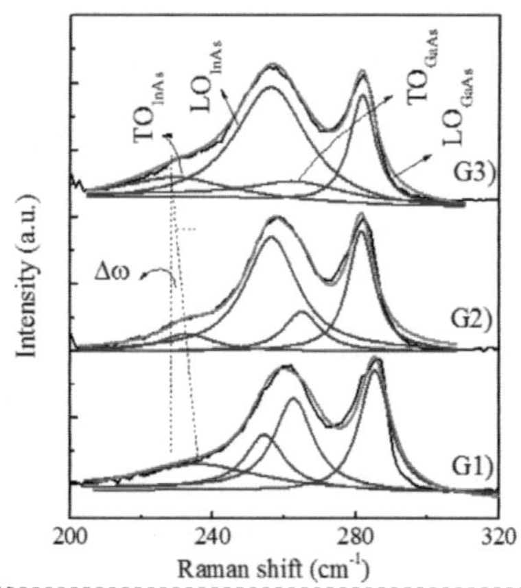

Figure 2. Raman spectra measurements of InGaAs thin films. The red line is the deconvolution of the experimental data using Lorentzians functions.

3.3. SEM

Figure 3 (a) shows a micrograph for InGaAs layer prepared with the lowest sputtering power. This sample presents a lowest roughness compared with the sample prepared to higher sputtering power [Figure. 3 b) and c)]. In Figure 3 (b), surface changes (some voids) are observed, due to increase in the lattice mismatch between layer and substrate, associated with increasing Indium content. Finally, in G3 layer, shown in Figure 3 (c), this phenomenon is more evident, the surface is more roughness that previous ones, due to the formation and agglomeration of particles (clusters) about 150 nm diameter.

Although a change in morphology of the layers with the sputtering power (or indium content) is observed, Indium segregation and secondary phase's formation are discarded.

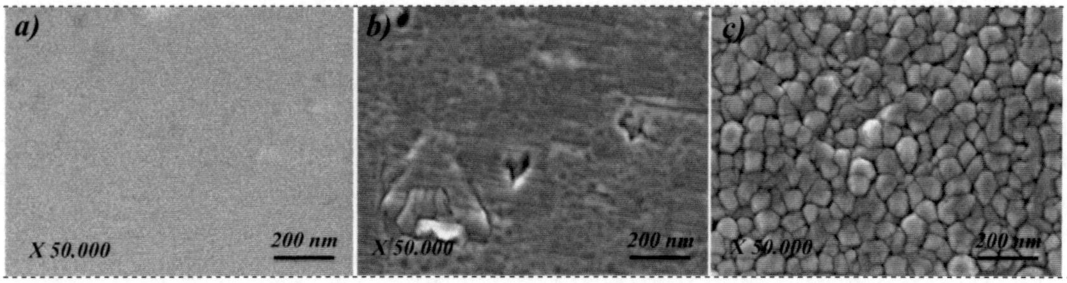

Figure 3. SEM Images on the surface of the InGaAs thin films deposited on a GaAs substrate at 4W (a), 6W (b) and 8W (c) Indium sputtering powers.

3.4. Compositional

3.4.1. SIMS. From SIMS depth profiling results, the distribution of the elements was determined. In Figure 4, we show only the spectrum for the InGaAs sample grown on a GaAs substrate at a 6W indium target power. The interface between the InGaAs layer and the substrate is clearly identified. The sample is homogeneous in composition because, the As, Ga and In signals intensity remains constant throughout the entire thickness. Traces of oxygen, nitrogen and carbon was not observed.

3.4.2. XPS. The Survey spectrum (Figure 5) corresponds to an InGaAs thin film in a region of 0-1200 eV. In this spectrum, characteristic states of InGaAs exposed to the environment, as well as Auger signals, characteristic of the ternary alloy, are observed. High-resolution measurements in the regions corresponding to Ga3d and As3d allowed the discretization on the behavior of the sample surface after being exposed to the environment. In said process, from a comparison of - In_2O_3 metal states, it can be demonstrated that As oxides are more volatile than Ga oxides. Spectra of the Ga were taken in the region between 14-24 eV (Figure 6). In this spectrum, from best fitting (blue line) on the experimental data the Ga hybridization states was identified.

In all spectra, a similar behavior is observed. However, only significant difference in the FWHM of In_4d state (narrow spectrum from 14 to 24 eV) is found.

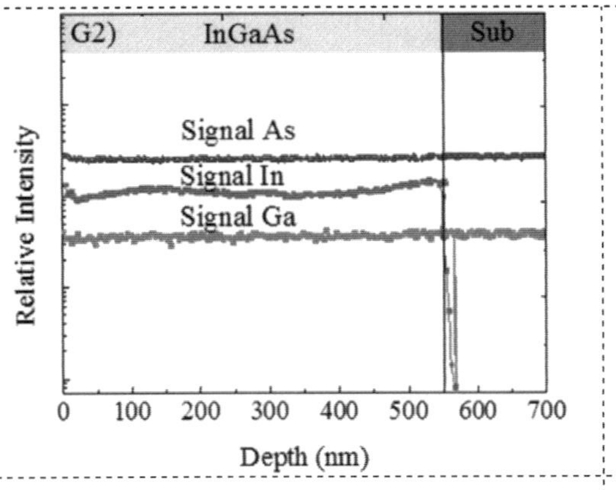

Figure 4. SIMS depth profiles of InGaAs layers of the sample prepared at a 6W indium target power.

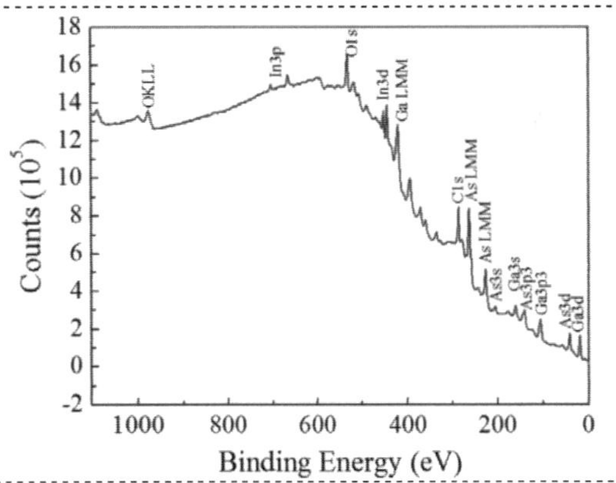

Figure 5. The Survey spectrum of the InGaAs semiconductor alloy.

The binding energies calculated for Ga_3d are in good agreement with that reported for the ternary semiconductor [13]. The present GaO is attributed to surface contamination post-process of sample preparation. The trace of GaO detected by XPS is due to the fact that it is a superficial analysis technique with a maximum penetration depth about 9 nm. In the same spectrum, a signal associated to In_4d state is shown, which depends on Indium concentration.

In the As spectra taken in 38-48 eV region (Figure 7), a similar analysis was performed. In this spectrum, the binding energy of As3d is also in good agreement with reported value for the ternary. Although hybridization states with oxygen As^{3+} are identified, the lowest levels of As oxidation are due to the high growth temperature, promoting either the formation or the not desorption of unstable oxides.

As for the states associated to Indium, in additional to the aforementioned 4d state, 3d state was observed in the narrow spectra taken in the 438-456 eV region. In each spectra, the In3d3/2 and In3d5/2 doublet was identified. In the layers deposited at Indium sputtering powers of 6 and 8 Watts (spectra not shown here), there are a change in its intensity, and FWHM of each of the signals, which is attributed to Indium concentration changes.

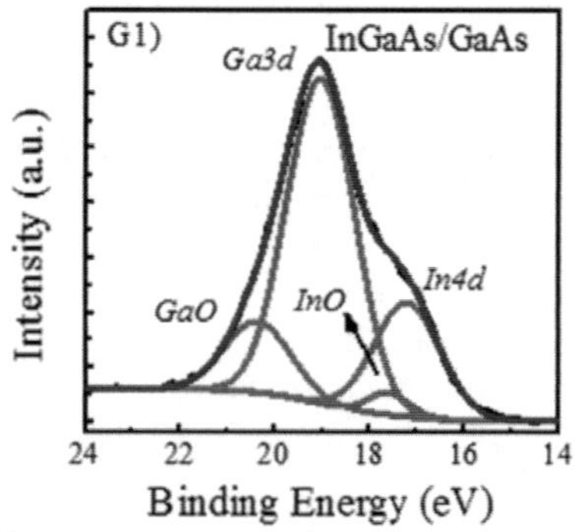

Figure 6. 14-24 eV region spectra of an InGaAs thin film on GaAs substrate (100) grown at Indium sputtering power of 6W..

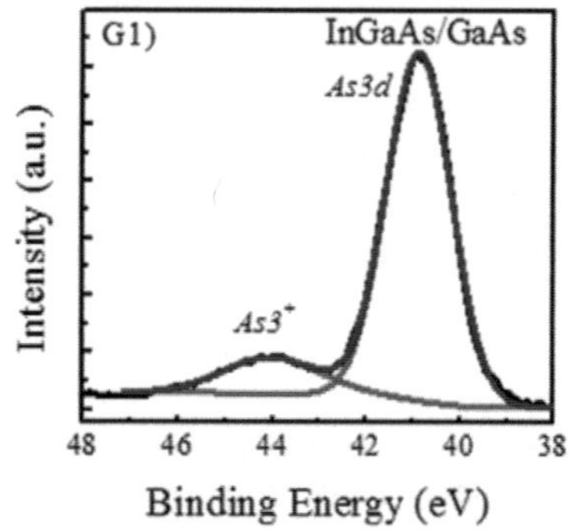

Figure 7. 38-48 eV region spectra of an InGaAs thin film on GaAs substrate (100) grown at Indium sputtering power of 6W.

4. Conclusions

It was found that the sputtering power allow varied the Indium content and/or the stoichiometry of the semiconductor at power lower than 8 Watts. Ternary InGaAs alloy present a preferential orientation along the direction (111), due to combined effects the stress and alloying. Finally, a dependence on its morphological properties as a function of the sputtering power was found.

Acknowledgments

Dirección de Investigaciones (DIMA) de la Universidad Nacional de Colombia - sede Manizales, proyectos 23196 y 30384. A. Tavira-Fuentes and A. Sotelo-García CINVESTAV-IPN México D. F. for their technical assistance. Roberto Bernal-Correa and Jorge Montes Monsalve, thanks COLCIENCIAS for scholarship "Francisco José de Caldas" – Call 528.

References

[1] Guo Q, Nakao T, Ushijima T, Shi W, Liu F, Saito K, Tanaka T and Nishio M 2014 J. Alloy. Compds. **587** 217–221
[2] Afzal N, Devarajan M and Ibrahim K 2016 Mater. Sci. Semicond Proce. **51** 8–14
[3] Venegas M, Bernal R, López M, and Pulzara A 2014 J. Phys: Conf. Series **480** 012017
[4] Boronat A, Silvestre S and Castañer L 2013 J. Non-Cryst. Solids **359** 21–26
[5] Lin K and Hau Kuo D 2016 J Mater Sci: Mater Electron **Article in press** 1-9.
[6] Kutz M 2015 Mechanical Engineers Handbook Manufacturing and Management (vol 3) ed Willey (United States of America) 245-247
[7] Habchi M, Tounsi N, Bedoui M, Zaied I, Rebey A, and El Jani B 2014 Phy. E **56** 74–78
[8] Shin H, Shin J, Do Yang K, Choi W and Choe J 2014 J. Nanoscience and Nanothecnology **14** 6297-6300
[9] Guo J, Yang M and Wang M 2013 Appl. Mech. Mater. **6** 49-442
[10] Pulzara A, Montes J, Bernal R, Morales A, Gallardo S, Lopez M 2016 Superficies y Vacio **29** 32-37
[11] Aslan M, Yalcın B, Üstündag M 2012 J. Alloy. Compds. **519** 55-59
[12] Iwami M, Watanebe Y, Kato H, Nakayama M and Sano M 1987 Thin Solid Films **3** 291-297

Ion hydration effects in the optical absorption and molecular vibration of water

Eduard Moreno, Janeth Fernandez P, Sandra M. Pinto and David A. Miranda

Universidad Industrial de Santander, Cra 27 Cll 9, Bucaramanga, Colombia

E-mail: dalemir@uis.edu.co

Abstract.
Hydration occurs when some ionic species are in pure water. In this research study, optical properties of an electrolytic solution with ionic species that emulated the extracellular environment were analyzed. The attenuated total reflection ATR-FTIR spectra and Raman spectra of electrolyte solutions were measured and the vibrational spectra of pure water (type A) with electrolyte solutions in the spectral region between 2700-4000 cm^{-1} was compared. At low ionic concentrations of chloride, sodium, potassium and calcium, a weak effect on the spectrum of water was observed. We found a shifts in the absorption bands of water in IR (3286 cm^{-1} and 3492 cm^{-1}) and Raman spectra (3207 cm^{-1} and 3432 cm^{-1}), which increase with ionic concentration. This effect can be interpreted as a result of interactions between water molecules and solvated ions. This result suggests that the presence of ions in pure water modifies the optical properties of water.

1. Introduction

Biological tissue has water in the Intra and extracellular media. The extracellular medium has some ions as chloride, sodium, potassium and calcium [1]. The optical properties of water in extracellular media in the range of 2700 to 4000 cm^{-1} is strongly affected by the existence of ions [2, 3, 4]. Each ion in electrolytic solution is surrounded by water molecules i.e. ions are hydrated, and different types of intermolecular interactions can occur [5, 6, 7]. Hitherto, the presence of some ions can change the natural vibrations of water, although these free ions do not have vibrational modes in the interest region [7, 8, 9].

Raman spectroscopy and ATR-FTIR are special techniques to measure and study molecular vibrations and optical absorption in different materials, including water [5, 10, 11]. We employed these two techniques to study the alterations in optical properties of the water when these ions are added. In this paper, we report the analyze of Raman and ATR-FTIR spectra of electrolytic solutions.

2. Materials and Methods

2.1. Ionic Solution Preparation

We used KCl, $CaCl_2$, and $NaCl$ salts to prepare the electrolyte solutions. Sodium chloride ($NaCl$) (5,84 mg), potassium chloride (KCl)(7,45 mg), and calcium chloride ($CaCl_2$) (14,7 mg) were added at room temperature in 25mL of Milli-Q Ultrapure water type 1. The ionic solutions were prepared under the values of level saturation for each salt. Different dilutions were arranged to study the ionic interaction with water.

2.2. FTIR and Raman measurement

The FTIR spectrometer Nicolet iS-50 was used to obtain the absorption (ATR) spectra. The interferometer mirror was moving at 0.6363 cm^{-1}/sec with a resolution of 8 cm^{-1} and 110 scans. The micro-Raman Horiba Evolution LabRam was used to obtain the Raman spectra. We used the excitation source of 532nm with 25mW, 10 seconds as exposure time, ten scans and a grid 600 gr/mm.

2.3. Data Analysis

After the data acquisition, the baseline correction was made with polynomial fitting. Subsequently, two Gaussian functions were used to fitting each band of spectra [12].

3. Results and discussions

The vibrational modes and optical absorption peaks of aqueous electrolytic solution have a slightly different behavior than water. To analyze how many vibrational and absorption peaks are present in Raman, and IR spectra of water, a decomposition of spectra in two or more peaks could be performed to obtain the best fit to experimental data [13, 14, 15]. In Figure 1 (a), (c) and (e) are shown the Raman spectra for pure water (in black) and four concentrations (4, 3, 2 and 1 M of $CaCl_2$, KCl, and NaCl) of the electrolytic solutions (in color), at room temperature. In Figure 1 (b), (d), and (f), two Gaussian functions were used to describe each experimental band. In a similar way, Figure 2 shown experimental IR spectra. The Gaussian function parameters values are in Tables 1 and 2.

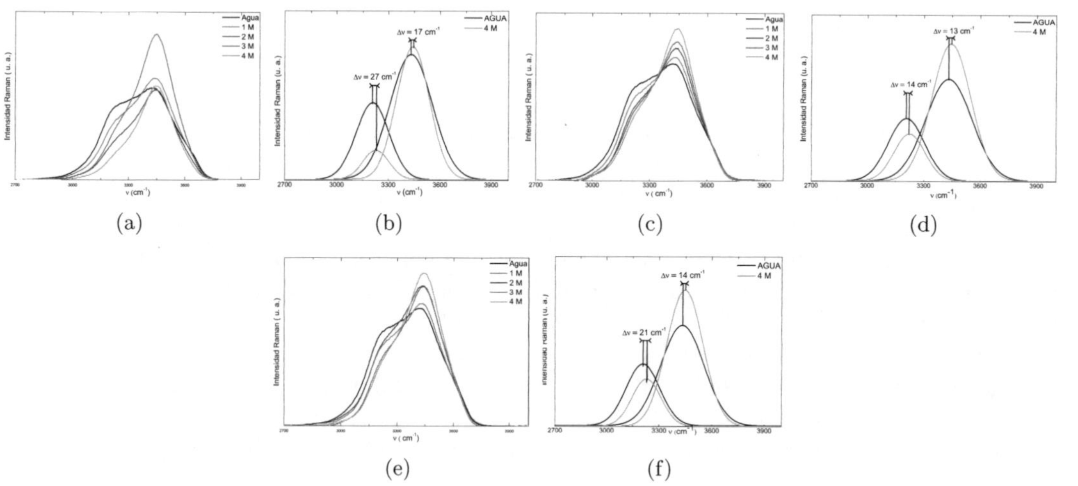

Figure 1. Raman spectra of water (in black) and aquous electrolytic solution (in color) of (a)$CaCl_2$, (c)KCl, (e) $NaCl$ and two-peak Gaussian fitting to (b) $CaCl_2$, (d) KCl, (f) $NaCl$.

Table 1. Gaussian function parameters values to Raman spectra of water and electrolytic solutions.

salt	Concentration (M)	Peak 1 (cm^{-1})	$\Delta\nu$ (cm^{-1})	Peak 2 (cm^{-1})	$\Delta\nu$ (cm^{-1})	COD R^2
Water	-	3207 ±1	-	3432 ±1	-	0,997
$CaCl_2$	1	3224 ±1	14	3448 ±1	16	0,998
	2	3227 ±1	17	3449 ±1	17	0,999
	3	3232 ±1	22	3449 ±1	17	0,999
	4	3234 ±1	27	3449 ±1	17	0,998
KCl	1	3207 ±1	0	3432 ±1	0	0,998
	2	3215 ±1	8	3438 ±1	6	0,999
	3	3219 ±1	12	3443 ±1	11	0,999
	4	3224 ±1	14	3445 ±1	13	0,998
$NaCl$	1	3214 ±1	7	3436 ±1	4	0,998
	2	3222 ±1	15	3444 ±1	12	0,998
	3	3223 ±1	16	3443 ±1	11	0,999
	4	3228 ±1	21	3446 ±1	14	0,999

The study of distances between ion and water molecules can be used to understand the ion-water interaction. The cations Na^{1+} and K^{1+} have a minor distance with the water molecules in the first shell of hydration, $3.3\mathring{A}$ and $3.7\mathring{A}$ respectively, whereas the distance of ion-water is $4.5\mathring{A}$ for Cl^{1-} [16]. The distance ion-water molecule influences the diffusion process of the ion through the cellular membrane. In particular, the concentration of ions sodium in the extracellular media increases according to the degree of malignancy of cervical cancer [17]. These differences between the Na^{1+}-water distances could explain why the membrane channel distinguishes explicit ions in solution such as Na^{1+} and why the shift frequency over Raman spectra is stronger for sodium more than potassium.

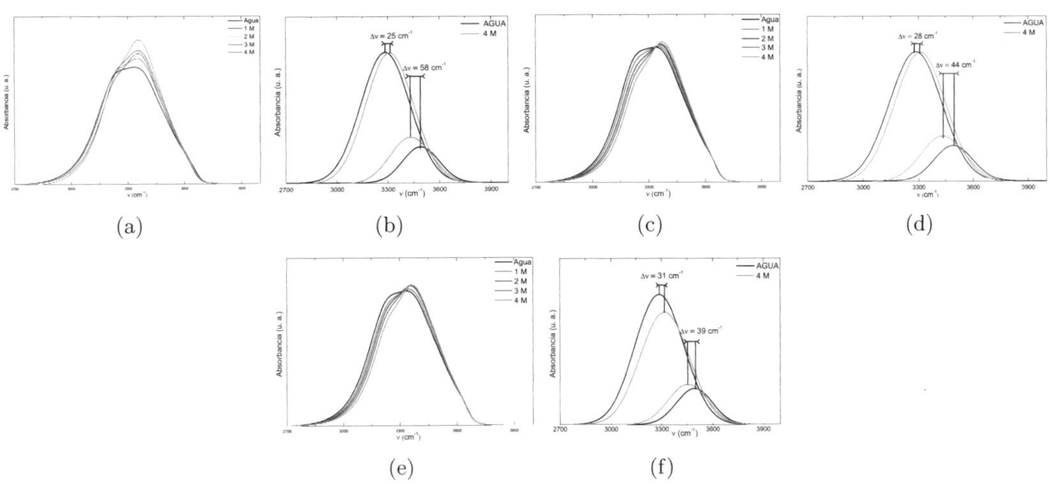

Figure 2. Absorption spectra of water (in black) and aquous electrolytic solution (in color) of (a)$CaCl_2$, (c)KCl, (e) $NaCl$ and two-peak Gaussian fitting to (b) $CaCl_2$, (d) KCl, (f) $NaCl$.

The two values of normal vibration of water obtained by optical absorption were $3286 \pm 8cm^{-1}$ and $3492 \pm 8cm^{-1}$, see Table 2. The experimental results to Na^{1+}, K^{1+} and Cl^{1-} are similar to reported previously i.e. these ions alter the normal frequency of water at high-frequency [4, 8, 10, 12, 14, 18, 19]. In this study, different concentrations were compared. The absorption spectrum of water for the natural vibration $3286 \pm 8cm^{-1}$ was shifted to the higher frequency and the $3492 \pm 8cm^{-1}$ natural vibrations, to the lower frequency with the increase of salts concentration.

Table 2. Gaussian function parameters values to absorption spectra of water and electrolytic solutions.

salt	Concentration (M)	Peak 1 (cm^{-1})	$\Delta\nu$ (cm^{-1})	Peak 2 (cm^{-1})	$\Delta\nu$ (cm^{-1})	COD R^2
Water	-	3286 ±8	-	3492 ±8	-	0,998
$CaCl_2$	1	3295 ±8	9	3479 ±8	13	0,998
	2	3301 ±8	15	3462 ±8	30	0,998
	3	3308 ±8	22	3443 ±8	49	0,998
	4	3311 ±8	25	3434 ±8	58	0,998
KCl	1	3297 ±8	11	3488 ±8	4	0,998
	2	3307 ±8	21	3479 ±8	13	0,998
	3	3311 ±8	25	3467 ±8	32	0,998
	4	3314 ±8	28	3448 ±8	44	0,998
$NaCl$	1	3298 ±8	12	3488 ±8	4	0,998
	2	3306 ±8	20	3480 ±8	12	0,998
	3	3313 ±8	27	3472 ±8	20	0,998
	4	3317 ±8	31	3453 ±8	39	0,998

4. Conclusions

The main ion hydration effect in the optical properties and molecular vibration is the shift in vibrational modes by the presence of ions. A substantial shift in the vibrational modes was observed with increasing salts concentration. The change in the frequency of vibration evidence ion induces alterations in the normal vibrations of the water since electrostatic interaction between ions and counterions in water molecules to affect the frequency of vibration. FTIR-ATR and Raman spectra show a similar effect. However, these spectra differ in the second sub-band, which corresponds to the antisymmetric stretching.

5. Acknowledgments

We are grateful to COLCIENCIAS by financial support and to the "Laboratorio de Espectroscopia (LEA)"at the "Universidad Industrial de Santander"by the measurement time.

References

[1] Hall J E 2010 *Guyton and Hall textbook of medical physiology* (Elsevier Health Sciences)
[2] Eisenberg D and Kauzmann W 2005 *The structure and properties of water* (Oxford University Press on Demand)
[3] Riemenschneider J 2011 *Spectroscopic investigations on pure water and aqueous salt solutions in the mid infrared region* Ph.D. thesis University of Rostock
[4] Sun Q 2012 *Vibrational Spectroscopy* **62** 110–114
[5] Wei Z F, Zhang Y H, Zhao L J, Liu J H and Li X H 2005 *The Journal of Physical Chemistry A* **109** 1337–1342
[6] Rull F 2002 *Pure and applied chemistry* **74** 1859–1870
[7] Walrafen G 1964 *The Journal of Chemical Physics* **40** 3249–3256
[8] Li F, Yuan J, Li D, Li S and Han Z 2015 *Journal of Molecular Structure* **1081** 38–43
[9] Irish D and Davis A 1968 *Canadian Journal of Chemistry* **46** 943–951
[10] Max J J and Chapados C 2001 *The Journal of Chemical Physics* **115** 2664–2675

[11] Max J J, Gessinger V, van Driessche C, Larouche P and Chapados C 2007 *The Journal of Chemical Physics* **126** 184507 URL http://scitation.aip.org/content/aip/journal/jcp/126/18/10.1063/1.2717184

[12] Sun Q, Zhao L, Li N and Liu J 2010 *Chemical Geology* **272** 55–61

[13] Marchal Y 2011 *Journal of Molecular Structure* **1004** 146 – 155 ISSN 0022-2860 URL http://www.sciencedirect.com/science/article/pii/S0022286011006247

[14] Duričkovć I, Marchetti M, Claverie R, Bourson P, Chassot J M and Fontana M D 2010 *Applied Spectroscopy* **64** 853–857

[15] Brooker M H, Hancock G, Rice B C and Shapter J 1989 *Journal of Raman Spectroscopy* **20** 683–694 ISSN 1097-4555 URL http://dx.doi.org/10.1002/jrs.1250201009

[16] Mancinelli R, Botti A, Bruni F, Ricci M A and Soper A K 2007 *The Journal of Physical Chemistry B* **111** 13570–13577 pMID: 17988114 (*Preprint* http://dx.doi.org/10.1021/jp075913v) URL http://dx.doi.org/10.1021/jp075913v

[17] Pinto S M, Corzo S P, Méndez-Sanchez S C and Miranda D A 2016 *Electrolytic Extracellular Phantom to Study the Low-Frequency Conductivity of Cervical Neoplasia* (Singapore: Springer Singapore) pp 9–11 ISBN 978-981-287-928-8

[18] Cappa C D, Smith J D, Wilson K R, Messer B M, Gilles M K, Cohen R C and Saykally R J 2005 *The Journal of Physical Chemistry B* **109** 7046–7052

[19] Kitadai N, Sawai T, Tonoue R, Nakashima S, Katsura M and Fukushi K 2014 *Journal of Solution Chemistry* **43** 1055–1077

IOP Conf. Series: Journal of Physics: Conf. Series **850** (2017) 012015 doi:10.1088/1742-6596/850/1/012015

Laboratory preparation questionnaires as a tool for the implementation of the Just in Time Teaching in the Physics I laboratories: Research training

David A. Miranda , Melba J. Sanchez and Oscar M. Forero
Universidad Industrial de Santander, Cr 27 Cll 9, Bucaramanga, Colombia
E-mail: dalemir@uis.edu.co

Abstract.
The implementation of the JiTT (Just in Time Teaching) strategy is presented to increase the previous preparation of students enrolled in the subject Physics Laboratory I offered at the Industrial University of Santander (UIS), Colombia. In this study, a laboratory preparation questionnaire (CPL) was applied as a tool for the implementation of JiTT combined with elements of mediated learning. It was found that the CPL allows to improve the students' experience regarding the preparation of the laboratory and the development of the experimental session. These questionnaires were implemented in an academic manager (Moodle) and a web application (lab.ciencias.uis.edu.co) was used to publish the contents essential for the preparation of the student before each practical session. The most significant result was that the students performed the experimental session with the basic knowledge to improve their learning experience.

1. Introduction

In the last years, the university education system has presented a change in its politics restructuring the traditional approach based on the passive transmission of knowledge, to move towards a comprehensive training approach that promotes competencies for life and contains multiple dimensions of knowledge [1]. In the year 2008, the Ministry of National Education of Colombia formulated the generic competences in higher education institutions.

For this reason, educations institutions have sought to incorporate Information and Communication Technologies (ICTs) into teaching-learning processes to improve the quality of teaching. However, despite the increased availability of technological resources in educational establishments, also, it has been shown that several teaching practices with ICT resources do not represent an advance, innovation or improvement of the traditional practices [2].

Several authors have reported their experiences with different pedagogical models for teaching science, one of which are inductive methods that include discovery-based, research-based, problem-based, project-based, case-based, and Teaching at the time, Just in Time Teaching (JiTT) [3, 4, 5]. A particular case of study is the JiTT proposed by Professor Gregor Novak. JiTT uses active learning and ICTs as the principal elements of the teaching strategy. Using this approach, Professor Novak observed a real change in the student's interest in learning physics [6] with a 40% reduction in students drop out in comparison with traditional methodologies

[7]. This strategy has also been implemented for the teaching of sciences such as chemistry, mathematics, economy, among others [8, 9].

The need for a research training strategy in the "Universidad Industrial de Santander" (UIS) was evident at 2013 because the self-assessment process performed to institutional accreditation. This process has shown the research training, given to bachelor students, as the lowest indicator in all the auto-evaluation process [10]. Another important factor that motivated the change in didactic strategy was the high index of desertion in the first levels. The development of investigative competencies implies the strengthening of skills to observe, ask experiment, interpret information, teamwork, appropriate use of technology, write reporting, order and systematize the investigative actions, publication, and knowledge management [11]. Because of this the Academic Vicerrectory, the Sciences Deanship and mainly the Physics School at UIS proposed a project in the year 2015 to improve the research training, with the objective to improve the education quality.

This study has the objective of innovating teaching and learning processes for the students enrolled in the Science and Engineering faculty programs through the Just in Time Teaching (JiTT) methodology to research training throw. The project started in the second academic semester of 2015 with the enrolment of all Physics I laboratory classes [12]. The methodology implemented encompasses three pedagogical strategies: active learning, mediated learning and just in time teaching in physics laboratories. In this paper, we describe the second implementation and some results of such strategies in the first semester of 2016.

2. Methodology

The implementation in the first academical semester of 2016 included 11 teachers and all the students taking Physics I laboratory class. Approximately 88% of the population (352 students) taking the class were surveyed. The Physics I laboratory course has a theoretical and an experimental component. Independent programs and teachers handle each of this components. The signature corresponds to the 20% of the class grading. This signature is graded as follows:

Table 1. Physics I laboratory grading.

Component	*Percentage*
Laboratory Preparation Questionnaires (CPL)	30%
Laboratory Reports (I)	40%
Final Project (Ω)	30%

The methodology consists of two parts, first out of the class (didactic) and second in the class.

2.1. Didactic or out of the class component

To prepare the practical sessions, a set of laboratory preparation questionnaires (CPL) were proposed to the students, where the JiTT strategy and mediated learning elements were implemented, all of this via the Moodle platform (tic.uis.edu.co) and the web app lab.ciencias.uis.edu.co for the content management. The CPL are a new tool that mixes the JiTT strategy with mediated learning based on ICTs. The JiTT consists of asking open questions to motivate student introspection. The mediated learning component consists of closed (mediation) questions related to the preparation of the laboratory. The mediation (closed) questions focus on the most important aspects to be taught during the lab while the JiTT questions aim at the identification of student argumentation. In particular, the closed questions were focused on equipment utilization and basic theoretical concepts and the JiTT questions in the experimental methodology to be implemented in the laboratory room. For the preparation of a practical session, each student was asked to review the published material in the web platform, which included projects, videos, equipment usage tutorials, and methodology related to the experiments carried out in the particular laboratory. The CPL were available 36 hours before the lab class in agreement with JiTT. The Moodle platform automatically graded the closed questions, and the teacher reviewed the JiTT questions.

2.2. In the class component

Based on the results of the out of the class component, the professor, at the beginning of the lessons, reinforced theoretical concepts that were found not to be adequately understood by the students, as the JiTT suggest. Following, the students proceed with the experiments more quickly given the previously (out of class) review of lab equipment, procedures, methodology, and other skill to perform their laboratory practice.

3. Measurement tool

At the end of the semester, an anonymous on-line survey for the students and professors was used to evaluate the impact of the strategies implementation.

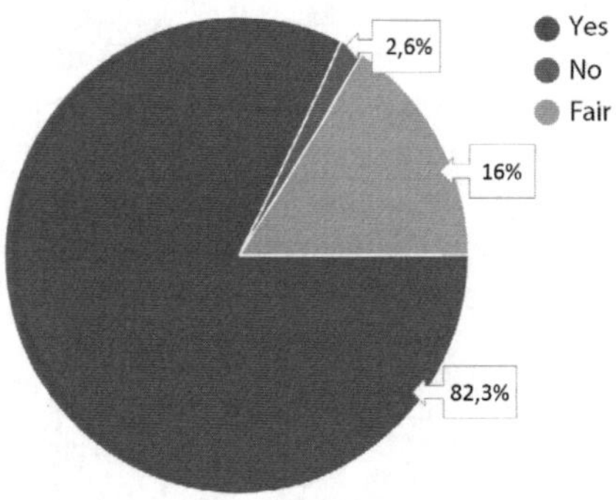

Figure 1. Summary of the question: Do you consider appropriate research projects?

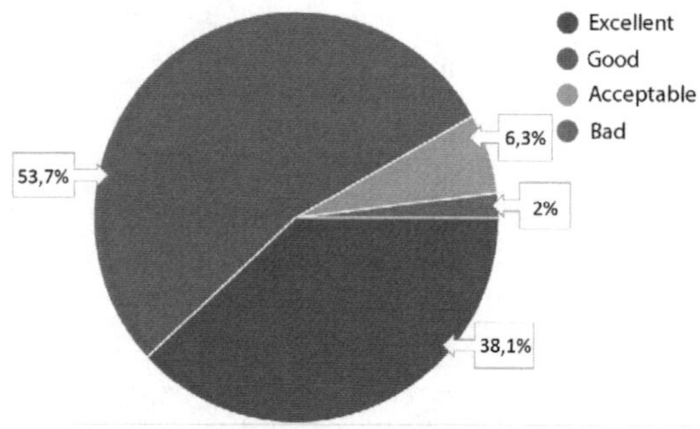

Figure 2. Summary of the question: How would you rank the use of Moodle platform in the laboratory?

3.1. Results

The survey at the end of the semester attended two purposes: first, to evaluate the student's acceptation level of the proposed methodology and second, to determine the improvement of the research training. The sample consisted of 88% of all the students taking the class and all the professors during this period. The results, collected via the Moodle platform, shown an improvement in the research training in the sense that student considered appropriated the research projects (figure 1) and the comments of the teachers with more than two years imparting the course. The acceptation of the methodology was very confident (figure 2). 91.8% of the students thought the platform useful for the strength hen of their research skills. Additionally, the majority of the students believed that the methodology optimizes the use of time allowing the execution of the experiments during the first two hours of the lab session (figure 3). The implementation of the CPL allowed identifying theoretical concepts not properly understood by the students; the teacher explained such concepts during the lab sessions. The professors noticed a real students change of attitude regarding the experimentation of theoretical concepts and more interested in its performing.

The use of open questions (JiTT) shown improvements related to deeper thinking in the students. However, some difficulties related to the preparation of lab reports and results interpretation were detected (figure 4). These problems were considered for the next phase of the project. Finally, related to the Moodle platform, problems were found related to students or theachers using it for the first time and repeated server failures. To solve the issues associated with the use of Moodle a training program was created by CEDEUIS. Regarding the server failures, the university will implement improvements to the ICT infrastructure.

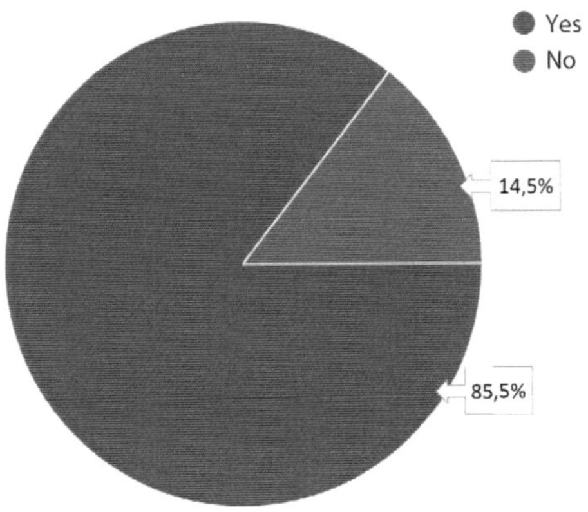

Figure 3. Summary of the question: Do you consider two hours is sufficient time for the lab session?

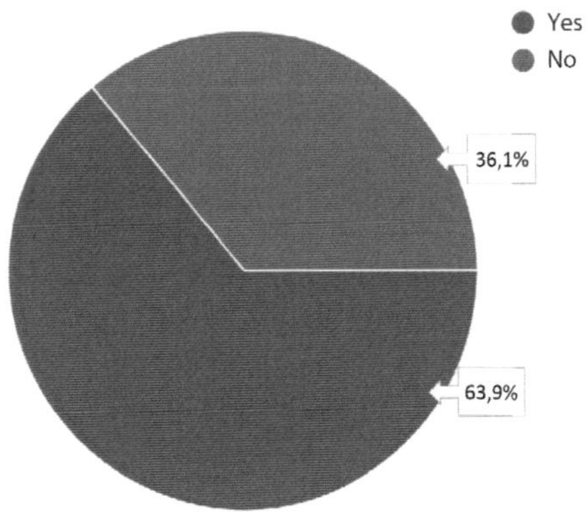

Figure 4. Summary of the question: After finishing the lab experiments, is it clear how to develop the lab report?

4. Conclusions

With the implementation of the methodology described in this paper, we observe an improvement in research training in the Physics I laboratories of the "Universidad Industrial de Santander". In particular, the development and implementation of the CPL allowed identifying theoretical concepts not properly understood by the students while preparing a lab session. In general, the methodology strengthened the teaching/learning process. However, issues related to data interpretation and results synthesis were found. These problems have been taking into consideration for the implementation of the methodology in the Physics Lab for the follow academical semester.

5. Acknowledgments

We are grateful to teachers and students of the laboratories of Physics at UIS that participate in the execution of the project.

References

[1] Trujillo F 2011 *"Enfoque de competencias en la educación: del conocimiento al uso y apropiación. Gestión de conocimiento"* (Ministerio de Educación Nacional, República de Colombia)
[2] Balanskat A, Blamire R and Kefalaq S 2006 The ICT Impact Report: A review of studies of ICT impact on schools in Europe Tech. rep. European Schoolnet URL `http://insight.eun.org/shared/data/pdf/impact_study.pdf`

[3] Draper A J 2004 *Journal of Chemical Education* **81** 221
[4] Kesner L and Eyring E M 1999 *Journal of Chemical Education* **76** 920
[5] Prince M and Felder R 2007 *National Science Teachers Association (NSTA)* URL `http://trove.nla.gov.au/work/16888534`
[6] Novak G 1999 *Just-in-time Teaching: Blending Active Learning with Web Technology* Ellis Horwood Series in Environmental Management, Science an (Prentice Hall) ISBN 9780130850348
[7] Marrs K A and Novak G 2004 *Cell Biol Educ* **3** 49–61
[8] Birk J P and Foster J 1993 *Journal of Chemical Education* **70** 180
[9] Hake R R 1998 *American Journal of Physics* **66** 64–74
[10] UIS 2013 *"Autoevaluación institucional 2013"* (Universidad Industrial de Santander) URL `https://www.uis.edu.co`
[11] Cota A 2011 *"Las competencias requeridas en investigacin y su grado de estmulo en Ingeniera mecnica del instituto tecnolgico superior de Cajeme"* (ITESCA) URL `http://www.itesca.edu.mx/investigacion/foro/carp ponencias/25.pdf`
[12] Miranda D A, Martínez J H, Sanchez-Soledad M J and Gómez-Bayona L 2016 *"Enfoque innovador de los laboratorios de física 1: formación para la investigación apoyada en TIC y estrategias de enseanza/aprendizaje"* (Universidad Industrial de Santander) URL `http://posgradoseeie.uis.edu.co/boletines/trabajoscogestec2016.pdf`

IOP Conf. Series: Journal of Physics: Conf. Series **850** (2017) 012016 doi:10.1088/1742-6596/850/1/012016

Legal Time of the Republic of Colombia and its international traceability using the Cesium Atomic Clock – Time and Frequency National Standard

Liz Catherine Hernández Forero and Nelson Bahamón Cortés

Instituto Nacional de Metrología de Colombia, Avenida Carrera 50 No. 26 – 55 Int. 2, Bogotá D.C., Colombia.

lhernandez@inm.gov.co, nbahamon@inm.gov.co

Abstract. Around the world, there are different providers of timestamp (mobile, radio or television operators, satellites of the GPS network, astronomical measurements, etc.), however, the source of the legal time for a country is either the national metrology institute or another designated laboratory. This activity requires a time standard based on an atomic time scale. The International Bureau of Weights and Measures (BIPM) calculates a weighted average of the time kept in more than 60 nations and produces a single international time scale, called Coordinated Universal Time (UTC). This article presents the current time scale that generates Legal Time for the Republic of Colombia produced by the Instituto Nacional de Metrología (INM) using the time and frequency national standard, a cesium atomic oscillator. It also illustrates how important it is for the academic, scientific and industrial communities, as well as the general public, to be synchronized with this time scale, which is traceable to the International System (SI) of units, through international comparisons that are made in real time.

1. Introduction

The routine of a human being in a social context entails the dating of events and their punctual execution, for example, the time of birth or death, the due dates for bills for public services or banking entities, the deadline for enrollment in a study program or for submitting a bid, the starting time of a sporting event, the schedule of working hours, the celebration of Christmas or the New Year, are just a few of the many events that require identifying a numerical reference of time.

According to the Spanish language dictionary Real Academia Española © [1], the meaning of punctuality is: "care and diligence in arriving to or leaving a place at the conventional time." Grammatically, the object of this sentence is "conventional time". As such, if we frame this object in a metrological definition, the key words are "time" and its adjective "conventional," specifically associated with the word "convention." That is to say, a time agreed upon or settled on by people, organizations, or countries. One of the characteristics of metrology is the linking of the units of measurements of reference, in particular, to the International System of Units (SI), which was established by the General Conference on Weights and Measures (CGPM), in turn created by the Metre Convention in 1875 [2]. Therefore, within the metrological framework of the SI, a relationship was established for dating events by using a numerical value for time linked to a convention on references, itself internationally identified as a time scale. The scale of time in a country is tied to legislation, or the set of rules or laws by which the State governs. As such, the point of interest in this document is to detail

IOP Conf. Series: Journal of Physics: Conf. Series **850** (2017) 012016 doi:10.1088/1742-6596/850/1/012016

the legal and metrological context for maintaining, coordinating, and disseminating Legal Time in the Republic of Colombia.

2. Metrological context of time

On June 22, 1799 in the National Archives in Paris, when the French Revolution was concluding, the initial step in developing the present SI was to create the decimal metric system and to define standards for the meter and kilogram [2]. With the signing of the international treaty called the "Metre Convention" on May 20, 1875, three inter-governmental organizations were created: the International Bureau of Weights and Measures, the General Conference on Weights and Measures, and the International Committee for Weights and Measures, whose acronyms in French are, respectively: BIPM, CGPM, and CIPM. The CGPM is comprised of delegates from every member state who meet every four years. As a result of each meeting, the CGPM makes decisions with regard to metrology, mainly associated with the International System of Units (SI). Each meeting is identified by a consecutive number. The content of the section of the present document on metrological context, section 2, uses as a reference the prescriptions of the eighth edition of the SI Brochure [2].

2.1. International system of units (SI)

The seven SI base units were established at the tenth CGPM in 1954 and at the fourteenth CGPM in 1971. The definition of each of these units is modified insomuch as scientific and investigative developments advance [2].

2.1.1. First definition of the second. The fraction 1/86 400 of the duration of a mean solar day: the exact definition of a "mean solar day" has an application for astronomers. It has been shown that this definition is inadequate due to irregularities in the earth's rotation. To have a more precise definition, during the eleventh CGPM in 1960, a definition provided by the International Astronomical Union was adopted based on the tropical year 1900 [2].

2.1.2. Second definition of the second. Experiments have shown that an atomic time standard based on the transition between two energy levels of an atom or a molecule can be produced and reproduced with much greater accuracy [2]. Taking into account that a precise definition of the unit of time is indispensable for science and technology, the thirteenth CGPM in 1967 replaced the definition of the unit for second with the following:

A second is the duration of 9 192 631 770 periods of the radiation corresponding to the transition between the two hyperfine levels of the ground state of the cesium 133 atom.

The 1997 CIPM meeting confirmed that this definition refers to a cesium atom at rest at a temperature of zero kelvin (0 K) [2].

2.2. Symbols, derived units and units outside the SI

Table 1 includes a summary of the units used for time and frequency, which are explained in sections 2.2.1 to 2.2.3.

2.2.1. Symbols. Writing the symbols of units correctly is mandatory. In the case of quantities of time, the symbol for the second in the SI is the letter "s" from the Latin alphabet. This is to say that the use of abbreviations such as "sec" or "sc" are not aligned with the stipulations of the SI.

Table 1. Units of time and frequency quantities.

Type of Unit	Quantity	Unit Name	Unit Symbol	Equivalences in terms of other units			
Basic	Time	second	s	1 s	0.017 min	0.00028 h	0.000012 d
Derived Coherent	Frequency	hertz	Hz	1 s^{-1}	Not apply.	Not apply.	Not apply.
Outside SI	Time	minute	min	60 s	1 min	0.017 h	0.00069 d
Outside SI	Time	hour	h	3 600 s	60 min	1 h	0.042 d
Outside SI	Time	day	d	86 400 s	1 440 min	24 h	1 d

2.2.2. Derived coherent unit. In the case of quantities of frequency, a coherent derived unit is used, given that the frequency is a power of the base unit of time that only includes a numerical factor equal to 1. This is to say that frequency is equal to the inverse of the period of oscillation (time). A special name and symbol was assigned to frequency: the name is "hertz" and the symbol is "Hz," which, when expressed in terms of the base units of the SI, is "s^{-1}" [2].

2.2.3. Units outside the SI. It is accepted that some units not belonging to the SI, such as the day, hour, and minute, still appear in scientific, technical, and commercial literature and will continue being used for the foreseeable future because they have historic significance and are profoundly rooted in the culture and daily life of the human race. Each one of these units has an equivalent in terms of the unit of the second. See Table 1 [2].

2.3. Time and frequency standards

For the metrological basics of time and frequency, it is highly important to identify the type of standard with which the unit of second can be reproduced. As is mentioned by Lombardi in [3]: "**all time and frequency standards** are based on a periodic event that repeats at a constant rate. The device that generates this event is called a **resonator**. In the simple example of pendulum clock, the pendulum is the resonator. Of course, a resonator requires an energy source to move backwards and forwards. Together, the energy source and resonator make up the **oscillator**. The oscillator operates at a rate known as the **resonance frequency**. For example, by design, a pendulum can swing once per second. The count of one complete swing of the pendulum produces a time interval of 1 s. By counting the total number of changes, a **time scale** is created which establishes the longest time intervals, such as minutes, hours, and days. The device that counts and visualizes or registers the results is called a clock".

2.3.1. Time bases. A time base is an oscillator within an electronic instrument. Its evolution is directly related to scientific advancements for achieving greater stability in a resonator, which results in a clock with greater accuracy. Table 2 provides a summary of the evolution of time and frequency standards and their uncertainty values (*U*) [3]. It is very important to elaborate on the order of magnitude of this evolution; from the 14th century to 1991 (approximately 700 years), the uncertainty of frequency started at 1×10^{-2} and evolved to 1×10^{-15}, that is to say, an improvement by 13 orders of magnitude. If we analyze the years 1921 to 1991 (approximately 100 years), the uncertainty of frequency started at 1×10^{-7}, meaning an improvement by 8 orders of magnitude.

In the majority of electronic instruments such as hand watches, computers, timers, cellular phones, global positioning systems (GPS), telecommunications equipment, among other appliances, the time bases used are quartz oscillators (hexagonal crystal prism) because of their application of the piezoelectric effect as a technological foundation. However, depending on the application and the context of required accuracy, some GPS or network synchronization instruments are manufactured with rubidium oscillators as their time base [4].

Table 2. Evolution of time and frequency standards [3].

Standard	Resonator	Date of Origin	U of time (24 h)	U of frequency (24 h)
Sundial	Apparent motion of the sun	3 500 B.C.	N/A	N/A
Verge escapement	Verge and Foliet mechanism	14th century	15 min	1×10^{-2}
Pendulum	Pendulum	1656	10 s	1×10^{-4}
Harrison chronometer (H4)	Spring and balance wheel	1759	350 ms	4×10^{-6}
Short pendulum	Two pendulums, slave and master	1921	10 ms	1×10^{-7}
Quartz crystal	Quartz crystal	1927	10 μs	1×10^{-10}
Rubidium gas cell	^{87}Rb resonance	1958	100 ns	1×10^{-12}
Cesium beam	^{133}Cs resonance	1952	1 ns	1×10^{-14}
Hydrogen maser	Hydrogen resonance	1960	1 ns	1×10^{-14}
Cesium fountain	^{133}Cs resonance	1991	100 ps	1×10^{-15}

2.3.2. Block diagram of a cesium atomic oscillator. Figure 1 shows a block diagram of a type of oscillator with a cesium beam, which is similar to the cesium oscillators that are sold commercially (cesium fountain devices are developed for research experiments in, for example, the United States and Germany). The internal operation of an atomic oscillator includes a quartz crystal; its position in the diagram is in the feedback loop. The theory on closed loop control systems [5] is used to explain the operation of the cesium oscillator with the objective of graphically representing the functions carried out by each component and representing the signal flow in figure 1. The definition of the functions is as follows:

- Process: Cesium beam technology and frequency standard outputs (5 MHz, 1 PPS).
- Reference: Resonance frequency of cesium atom (^{133}Cs), which is equal to 9 192 631 770 Hz.
- Sensor: Detects atoms whose energy status has changed while passing through the microwave oven cavity (vacuum cavity, getter and detector).
- Controller: Must continuously synchronize the quartz oscillator and synthesize resonance frequency (frequency synthesizer, quartz oscillator and servo feedback).
- Actuator: Cesium gas oven, status detection magnets and microwave oven cavity (cesium oven, state selection magnets and microwave interrogation cavity).

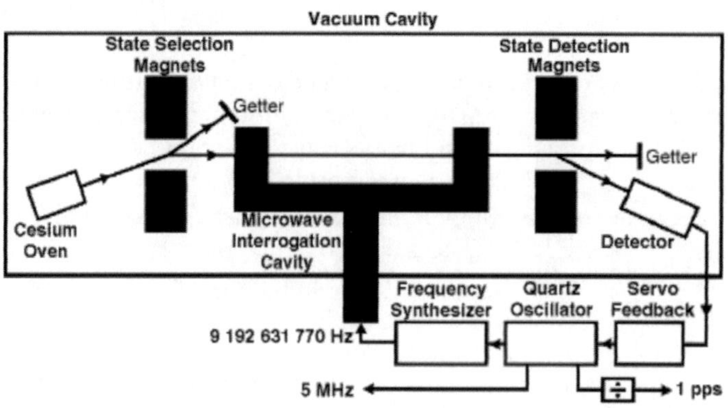

Figure 1. Block diagram of the cesium beam oscillator [3].

An explanation of the process of a cesium beam oscillator is provided by Lombardi in [3]: "The ^{133}Cs gas atoms are heated in an oven. Atoms from the gas leave the oven in a high-velocity beam that travels through a vacuum tube toward a pair of magnets. The magnets serve as a gate that allows only atoms of a particular magnetic energy state to pass into a microwave cavity, where they are exposed to a microwave frequency derived from a quartz oscillator. If the microwave frequency matches the resonance frequency of cesium, the cesium atoms change their magnetic energy state. The atomic beam then passes through another magnetic gate near the end of the tube. Those atoms that changed their energy state while passing through the microwave cavity are allowed to proceed to a detector at the end of the tube. Atoms that did not change state are deflected away from the detector. The detector produces a feedback signal that continually tunes the quartz oscillator in a way that maximizes the number of state changes so that the greatest number of atoms reaches the detector. Standard output frequencies are derived from the locked quartz oscillator".

2.4. Time scales

A time scale depends on the oscillator of reference (time base) and the specific objective of the scientific or legal application. The time base of each scale has evolved from astronomical measures with reference to celestial bodies such as the moon (ephemeris time), the stars (sidereal time), the sun (solar time, Julian date, Greenwich time, universal time (UT) and adjustments to the UT to minimize the effect of inhomogeneities in the earth's rotation), even the current atomic scale (TAI, UTC). Table 3 has a summary of the main time scales.

2.4.1. UTC time scale. UTC is the atomic time scale that forms the base for the coordinated diffusion of standard frequencies and time signals [6]; it is generated and monitored by the BIPM. The importance of each country generating its own atomic scale is that the UTC is an average weighted from these atomic scales. This means that the indication of international time associated to the UTC is not generated by one cesium clock alone, as described in Section 2.3.2, but rather that the UTC is a scale derived from a mathematical algorithm and not from a physical clock.

Table 3. Evolution of time scales.

Time scale	Reference	Purpose
Solar time	Apparent motion of the sun over the horizon of a place.	Base of all calendars. The highest point in the sky signifies "noon".
Greenwich Time (GMT)	Mean solar day as observed at the Greenwich Meridian.	Definition of time zone. A count from 12:00 one day to 12:00 the following day.
Universal Time (UT)	GMT	A count from 00:00 one day to 00:00 the next day, in 1925.
Universal Time 0 (UT0)	1 tropical year (vernal point) = 365 d + 5 h + 48 min + 45.5 s	UT0 = Minimize the effects of the eccentricity and inclination of Earth's orbit.
Universal Time 1 (UT1)	1 second = 1 mean solar day / 86 400	UT1 = UT0 + True angular rotation movement of Earth.
International Atomic Time (TAI)	Readings from different atomic clocks.	Define a coordinated time reference in 1970 and adjusted in 1980.
Coordinated Universal Time (UTC)	Leap Seconds (N) are applied: UTC = TAI – N [7].	In the first half of 2017: N = 37 leap seconds.

Countries that participate in the creation of the UTC are those countries who have signed the "Metre Convention," the treaty that is now known as the CIPM MRA – CIPM Mutual Recognition Arrangement. It is important to emphasize that the CIPM MRA is the framework through which national metrology institutes (NMIs) can show international equivalence of their measurement standards and the calibration and measurement certificates that they issue. Proof of this international equivalence is realized through peer reviews and through the approval of CMC – Calibration and Measurement Capabilities. These CMC are published, so as to be available to the public, in the CIPM MRA database, called the "The BIPM key comparison database" (KCDB).

In 1875, there were 17 initial member countries in the Metre Convention. Currently (2016), the CIPM MRA has been signed by representatives of 102 institutes, distributed in 57 Member States, 41 Associates of the CGPM, and 4 international organizations. In addition, it covers 153 other institutes designated by the signatory bodies [6]. This implies that the CIPM MRA provides the technical base for the negotiation of international commerce and regulatory matters in every country in the world.

This is how the BIPM creates two technical bases. First, it creates a scale called TAI – International Atomic Time, which is the strict average weighing the time scales of all signatory countries who send data from their clock signal to the BIPM, located in Sèvres, France. The TAI scale should be adjusted to ensure a scale of reference that is approximately in agreement with the time derived from the earth's rotation; the adjusted TAI is the aforementioned UTC.

3. Legal context of time

Every country establishes its use of a system of units for general purposes or for use in specific areas such as for commerce, health, public security, and education. In the specific case of the quantities of time and frequency, the technical base corresponds to the numerical indication of the legal time for the country and for the entity designated for disseminating that indication.

3.1. International context

Using for reference the countries that are currently member states in the CIPM MRA, Table 4 includes some examples of institutes whose function is broadcasting the legal time of the country and the website where the time is published for the citizenry.

Table 4. Examples of Institutes that disseminate the legal time of their country.

Country	Entity	Type of institute	Website with the legal time
Colombia	Instituto Nacional de Metrología (INM)	National	horalegal.inm.gov.co
Germany	Physikalisch-Technische Bundesanstalt (PTB)	National	ptbtime1.ptb.de
Costa Rica	Instituto Costarricense de Electricidad (ICE)	Designated	www.grupoice.com/wps/portal/ICE/AcercadelGrupoICE/Laboratorios/LMVE
United States	National Institute of Standards and Technology (NIST)	National	nist.time.gov
Italy	L'Istituto nazionale di ricerca metrologica (INRiM)	National	www.inrim.it/ntp/webclock_i.shtml
México	Centro Nacional de Metrología (CENAM)	National	www.cenam.mx/hora_oficial

3.2. National context

Documents and Administrative Acts from the Colombian government are a tool for establishing the legal parameters that govern the country's citizenry. Table 5 is a summary of the implementation history of the international system of units, of the definition of legal time, and of the designation of national measurement standards. It is important to detail two aspects:

- Regarding the declaration of the national time standard: the difference between the resolutions issued in 2011 and 2013 stems from extricating a specific instrument brand, model, and serial number from the declaration, given that commercial cesium oscillators generally have a useful life span of 5 to 12 years. Therefore, it is reasonable that the declaration should mention the type of atomic isotope of cesium used to construct the primary standard clocks, which by the SI definition of the second, is isotope 133 (see Section 2.1.2).
- The current adoption of the UTC reduced by 5 hours as the legal time in the Republic of Colombia is directly related to the geographic location of the country. Colombia is located in the fifth time zone to the left of the Greenwich meridian (see Figure 3).

Table 5. State of the art related to the legislation on the Legal Time in the Republic of Colombia.

Year	Administrative Act	Issuing entity	Content
1853	Ley 8	Senado y cámara de representantes de la Nueva Granada.	Legal text adopting the French decimal metric system for all purposes in Colombia [8].
1982	Decreto 2707	Ministerio de comunicaciones	Legal time = UTC – 5 hours. Broadcasted by the Centro de Control de Calidad y Metrología of the Superintendencia de Industria y Comercio (SIC).
1992	Decreto 717	Ministerio de Desarrollo Económico	Legal time = UTC – 4 hours. Purpose: To achieve electric power saving through the use of sunlight. Broadcasted from 1992-05-01 until 1993-02-07.
2011	Decreto 4175	Ministerio de Comercio, Industria y Turismo (MINCIT)	The Instituto Nacional de Metrología (INM) is created. One of its functions: "Maintain, coordinate and broadcast the legal time of the Republic of Colombia".
2011	Resolución 6467	MINCIT and SIC	The national time standard is declared to be the following: Cesium atomic clock, Brand: Symmetricom, Model: Cs III 4310 B, Serial number: 0833008915.
2013	Resolución 41242	MINCIT and SIC	The national time standard is declared as follows: Name: Atomic clock. Description: set of cesium 133 atomic clocks.
2015	Decreto 1595	MINCIT	The businesses, entities, and bodies dedicated, among its services, to informing or using the legal time in some way, should disclose the legal time coordinated by the INM.

3.3. Written format

In the international context, the standard of reference is ISO 8601:2004, titled: "Data elements and interchange formats - Information interchange - Representation of dates and times". In Colombia this standard was adopted through Colombian technical standard NTC 1034:2014, titled "Elementos de datos y formatos de intercambio - Intercambio de información - Representación de fechas y horas". The former version of this standard was issued in 1994. The two main characteristics are:

- **Day of the year:** to write the time periods of day, month, and year numerically, the order used should be the time period which descends from the greatest to least duration and be separated by hyphens (year-month-day).
  Example: the date September 14, 2016 should be written as 2016-09-14.
- **Time of day:** to write the hour by taking into account the three elements that indicate the accumulation of seconds in a day—hours, minutes, and seconds— NTC 1034:2014 indicates that two digits should be written for each of the three elements, separated by the colon symbol and, that the 24-hour format be used, which is to say that the hour of the day be counted consecutively from the day's start to end, where the day's end coincides with the start of the next day. This moment is commonly known as "midnight". Due to the sexagesimal system, the count of the quantity of seconds in a minute and the quantity of minutes in an hour is an interval of 00 to 59 and the quantity of hours in a day is interval of 00 to 23.
  Example: the 57th second of the 19th minute of the 7th hour is 07:19:57.

4. Metrological traceability

The INM establishes traceability of its own measurement standards to the SI by means of an uninterrupted chain of calibrations tying them to the definition of the respective primary standards of SI units of measurement. This link is accomplished by referring to the national measurement standards or to reference standards calibrated by other metrology institutes recognized in the MRA. In the case of quantities of time and frequency, the national standard corresponds to a primary standard. As such, the cesium atomic oscillator designated as the national standard does not take into account the issuance of calibration certificates. Instead, its metrological assurance depends on international comparisons tying its traceability to the UTC time scale. Remembering what was described in Section 2.4.1 about the UTC time scale, it can be said that the BIPM maintains TAI and UTC time scales "on paper" [3]. Therefore, in order to estimate the uncertainty of the measurement processes of the Time and Frequency Laboratory at the INM, there are two international comparison options (see 4.1 and 4.2) that are executed in real time thanks to satellite communication.

4.1. INM traceability through the SIM

As is mentioned in Lopez et al [9], the Interamerican Metrology System Time Network (SIMTN) began operating in 2005 and continuously compares the time standards of the laboratories belonging to the SIM network. In 2011 there were 16 laboratories and currently there are 25 under comparison. The SIMTN reports results via the internet in real time every 10 minutes (http://tf.nist.gov/sim), making it easy to identify fluctuations in frequency and time in the short term and making possible the resolution of measurement problems. Since 2008, SIMTN data has been used to generate the SIM time scale (SIMT). Since the second half of 2009, the SIMT has been automatically generated with the results published in real time via the internet. Each time scale is denominated as SIMT(k) or UTC(k), with k as the acronym for the laboratory of the country participating in the comparisons. In Colombia's case, k is the INM. With the SIMTN system, the behavior of the local time scale of Colombia SIMT(INM) is registered, making it possible to estimate two parameters, the frequency offset, $(\boldsymbol{\Delta f/f})$, as well as its standard uncertainty, $\boldsymbol{u(\Delta f/f)}$. The final result is an estimation of the best Calibration and Measurement Capabilities (CMC) of the INM, whose declared value is 6.0×10^{-13} [10]; during the estimation process, the time scale of the United States' NIST was used for reference. Thus, the time differences of interest are SIMT(NIST) – SIMT(INM), given that SIMT(NIST) is the most stable time scale of the participants in the SIMT scale and, of the most importance metrologically-speaking, it has traceability to the UTC through the UTC – UTC(NIST) time differences. Note that the declared value is between the two values of uncertainty in frequency of the rubidium and cesium beam from Table 2.

In the months of June and July of 2016, the contribution percentage to the SIMT(INM) was a maximum value of 4 % and in the months of August and September the contribution percentage increased to approximately 9 %. Meanwhile, the United States has maintained a percentage of support between 35 % and 40 % from June to September, thus showing that the SIMT(NIST) is a scale of great stability. In this period of time, the numerical value of the frequency offset parameter of the time differences SIMT (NIST) - SIMT (INM) was equal to 1.39×10^{-14}, which is directly associated with the accuracy of the clock with which the legal time of Colombia is produced, thus, the hour produced by the INM would advance one second with respect to the hour produced by the NIST in a period of time approximately equal to 2 million years; with the metrological assurance performed by the INM, this frequency offset value can be reduced up to about $\pm 7.0 \times 10^{-15}$, thus, the period of time could be extended up to about 5 million years to get to visualize the advance (due to the positive sign) or delay (due to the negative sign) of one second. If this type of comparison is performed using the time produced by a stopwatch made with quartz oscillator against the legal time, then the frequency offset would be approximately equal to a value between $\pm 1.0 \times 10^{-5}$ to $\pm 1.0 \times 10^{-6}$, which means that the time produced by the stopwatch would be advanced or delayed by one second with respect to the time produced by the INM in a period of approximately 1 day to 11 days.

4.2. Future work to contribute to the realization of the UTC

As a research project, the INM is studying the methods that facilitate the completion of two goals: increasing its percentage contribution to the SIMT scale and stably maintaining this percentage. Of the percentages described in Section 4.1, it is evident that the first goal has already been reached, using an objective framework of a contribution percentage of between 8 % and 10 %. However, it is necessary to maintain the stability of this percentage during a period of at least one hundred consecutive days. This would constitute the first step for sending data from the INM clock signal to the BIPM. The final objective is to have the INM time scale included in the monthly publication called Circular T [6], which provides traceability to the UTC and with which the contribution percentage to the realization of the UTC through the time differences identified as UTC – UTC(INM) are documented.

4.3. Traceability of Colombian citizens with the INM

The legal time is published for all citizens at URL http://horalegal.inm.gov.co/. Businesses and entities can synchronize their equipment with the INM server. The same can be done with personal computers and devices through different applications that are available; more details can be found on the webpage: http://www.inm.gov.co/index.php/servicios-inm/hora-legal. The services provided by the Time and Frequency Laboratory can be consulted in the link: http://www.inm.gov.co/images/Docs/CMCs/Tiempo_y_Frecuencia_Espanol_2016-v1.pdf. Figure 2 shows the traceability chain.

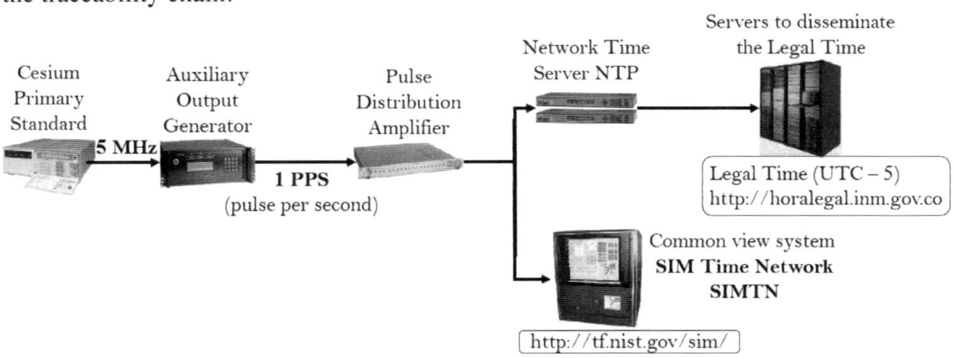

Figure 2. Traceability chain of the legal time of the INM time and frequency laboratory.

The INM systems technology department can monitor the different users that connect to reference the time by using a web analysis tool called "Google Analytics". The global information is summarized in Figure 3. As demonstrated, almost all of the countries in the world have connected at some time to reference the legal time in Colombia. This information is from a one-year period (Figure 3 shows specifically the log-on sessions from Russia). Table 6 shows the ten countries with the greatest number of log-ons to consult the legal time in the last year.

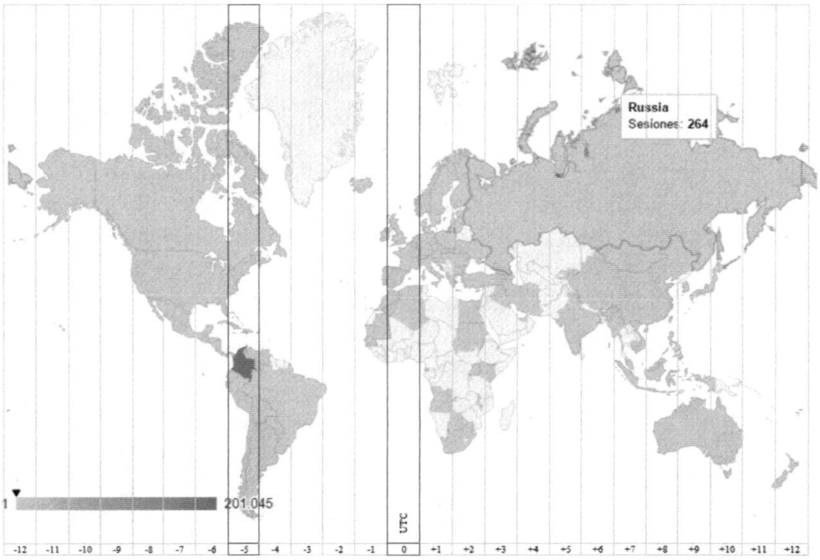

Figure 3. Users of the legal time issued by the INM and geographic location of UTC – 5 hours.

Table 6. Quantity of log-ons of ten countries.

Country	Quantity of log-ons	Percentage of consultations
Colombia	201 045	90.33 %
Spain	7 128	3.20 %
Venezuela	1 891	0.85 %
Chile	1 849	0.83 %
United States	1 402	0.63 %
France	915	0.41 %
Argentina	813	0.37 %
Germany	614	0.28 %
Italy	574	0.26 %
Panamá	542	0.24 %
Total	216 773	97.4 %

5. Examples of the use of time

The INM uses a 24-hour format to issue the legal time; the advantage of this is avoiding ambiguity. It is common to ask if 12 p.m. is 12 noon or midnight. The answer can be found at: http://www.inm.gov.co/index.php/serviciociudadano/informacion-para-el-ciudadano/preguntas-frecuentes

6. Conclusions

Just as every country has its national anthem, every country designates which entity in the national territory provides the time used for activities associated with legal compliance. The Legal Time of the Republic of Colombia depends on a cesium oscillator which is a primary standard because the SI second is defined according to the resonance frequency of the 133 cesium atom. This oscillator was designated as a national standard by means of an Administrative Act of the Colombian government. Metrological assurance results from comparisons of real time across different time scales: SIMT(NIST) – SIMT(INM) and UTC – UTC(NIST). The numerical result of these comparisons indicates that legal time is produced with an accuracy level of 1 second of advance or delay that would be displayed in a period of time ranging from 2 million to 5 million years, thus evidencing the reliability of synchronizing any clock with the Legal Hour broadcasted by the INM.

References

[1] Real Academia Española © (RAE) 2014 *Spanish Language Dictionary 23rd edition Tercentenary Edition* [Consulted in August 2016] Available at: http://dle.rae.es/

[2] Bureau international des poids et mesures BIPM 2006 *SI Brochure: The International System of Units (SI)* (Paris: STEDI Media) **8** pp 95-177

[3] Lombardi M 2002 *Chapter 17, Fundamentals of Time and Frequency (The Mechatronics Handbook)* (Austin, Texas: CRC Press LLC) pp 341-358

[4] National Institute of Standards and Technology (NIST) Time and Frequency Division, Popular links 2015 *Time and Frequency from A to Z: An illustrated glossary* [Consulted in August 2016] Available at: http://www.nist.gov/pml/div688/grp40/glossary.cfm

[5] Ogata K 2010 *Ingeniería de control moderna* (Madrid, España: PEARSON) pp 17-22

[6] Bureau international des poids et mesures BIPM 2016 *International reference time scales, BIPM services* [Consulted in September 2016] Available at: http://www.bipm.org/en/bipm-services/timescales/
International equivalence of measurements : the CIPM MRA, International equivalences [Consulted in September 2016] Available at: http://www.bipm.org/en/cipm-mra/

[7] International Earth Rotation and Reference Systems Service IERS *Bulletin C 52 Observatoire de paris* [Consulted in September 2016] Available at: ftp://hpiers.obspm.fr/iers/bul/bulc/bulletinc.dat

[8] Banco de la República 1853 and Biblioteca Luis Ángel Arango *Virtual library Miscellany 635* [Consulted in September 2016] Available at: http://www.banrepcultural.org/blaavirtual/derecho/lei-de-8-de-junio-de-1853

[9] Lopez J, Lombardi M, Novick A, Jimenez F, Diaz N, Lopez E, Boulanger J, Pelletier R, Carvalho R, Solis R, Sanchez H, Hernández L, Gordon J, Perez D, Bances E, Trigo L, Masi V, Postigo H, Questelles A, Gittens A and Ramos R 2011 *Real-Time Generation of the SIM Time Scale (SIMT) from International Clock Comparisons Proc. XXXth Gen. Assem. URSI.* p 4

[10] Hernández L and Chaparro G 2014 *Trazabilidad en tiempo y frecuencia usando método de calibración por diferencia de fase* (*Memorias de Simposio de Metrología 2014*) (Querétaro, México: Centro Nacional de Metrología CENAM) ISBN 978-607-96162-9-8 pp 681-686

IOP Conf. Series: Journal of Physics: Conf. Series **850** (2017) 012017 doi:10.1088/1742-6596/850/1/012017

Localized Plasmon resonance in metal nanoparticles using Mie theory

J.S. Duque[1], J.S. Blandón[1], H. Riascos[1].

[1] Grupo Plasma, Láser y Aplicaciones, Universidad Tecnológica de Pereira, Colombia.

E-mail: joseduque@utp.edu.co

Abstract. In this work, scattering light by colloidal metal nanoparticles with spherical shape was studied. Optical properties such as diffusion efficiencies of extinction and absorption Q_{ext} and Q_{abs} were calculated using Mie theory. We employed a MATLAB program to calculate the Mie efficiencies and the radial dependence of electric field intensities emitted for colloidal metal nanoparticles (MNPs). By UV-Vis spectroscopy we have determined the LSPR for Cu nanoparticles (CuNPs), Ni nanoparticles (NiNPs) and Co nanoparticles (CoNPs) grown by laser ablation technique. The peaks of resonances appear in 590nm, 384nm and 350nm for CuNPs, NiNPs and CoNPs respectively suspended in water. Changing the medium to acetone and ethanol we observed a shift of the resonance peaks, these values were agree with our simulations results.

1. Introduction

The optical properties of nanostructures, especially metal nanoparticles have taken great relevance due to many applications in catalysis process, plasmonics, optoelectronics, sensors and medicine. Nanoparticles are very interesting since show exotic properties such as the electric field confinement and Localized Surface Plasmons Resonances (LSPR) the latter is a phenomena characteristic of metallic nanoparticles, which is produced by charge oscillation in the surface of NP's in presence of an electric field. In spherical metal nanoparticles these are known as Fröhlich resonances (Bohren and Huffman 1983) [1, 2], and are the fundamental electric dipole resonances of spheres [3, 4]. In recent years several authors have investigated the LSPR characteristics for metal nanoparticles with spherical shape using simulations based on the Mie theory, in order to understand the behaviour of the optical properties of MNP's when changing the liquid environment or some of the parameters of growth, like show Hossein and collaborators in your work [5].

There are many techniques to obtained colloidal MPNs, including chemical and physical methods [6], in this paper MNPs were fabricated by laser ablation technique, this method of synthesis is simple and relatively fast, so that nanoparticles can be synthesized with a variety of forms and chemical natures [7]. By UV-Vis absorbance spectra we can to determine the regions where the radiation absorption has a maximum for colloidal metal nanoparticles grown by laser ablation with spherical shape. The electric filed intensities emitted for MNPs of different sizes were measured by nanoparticle tracking analysis (NTA) method, in each kind of MNPs. The electric field intensities of

MNPs too were analysed by simulation of Mie theory, also using this theory we could found the relationship of the peaks present in the scattering spectra with the LSPR for CuNPs, NiNPs and CoNPs. The dependence of the peaks shifted with the liquid environment change was determined.

2. Theory

Gustav Mie (1908) in his most popular work *"contributions to the optics of diffuse media, especially colloid metal solutions"* solved Maxwell's equations for the case of an incoming wave interacting with a spherical colloidal particle [1]. For the theoretical solution he considered a plane electromagnetic wave hits a smooth spherical particle embedded in a liquid medium with a real refractive index. The particle can be metallic, thus it can be absorbing and its refractive index is, in general, a complex number. In essence, the electromagnetic fields are expanded in multipolar contributions, and the expansion coefficients are found by applying the correct boundary conditions for electromagnetic fields at the interface between the metal nanoparticles and its surrounding medium. Mie introduced spherical coordinates system, used the dimensionless size parameter $x = 2\pi mR/\lambda_0$, where R is the particle radio, m is the real refractive index and λ_o is the incident wavelength. For the solutions of electromagnetic fields Mie introduced the extinction, absorption and scattering diffusion efficiencies, Q_{ext}, Q_{abs} and Q_{sca} known as Mie efficiencies of a spherical NP with area πR^2, this efficiencies are the cross section of extinction, absorption and scattering normalized per area unit and it is given by the following expressions [5, 8]:

$$Q_{ext} = \frac{\sigma_{ext}}{\pi R^2} = \frac{2}{x^2} \sum_{n=0}^{\infty}(2n+1)\Re(a_n + b_n) \quad (1)$$

$$Q_{sca} = \frac{\sigma_{sca}}{\pi R^2} = \frac{2}{x^2} \sum_{n=0}^{\infty}(2n+1)(|a_n| + |b_n|) \quad (2)$$

$$Q_{abs} = Q_{ext} - Q_{sca} \quad (3)$$

$$a_n = \frac{\psi_n(x)\psi'_n(mx) - m\psi_n(mx)\psi'_n(x)}{\xi_n(x)\psi'_n(mx) - m\psi_n(mx)\xi'_n(x)} \quad (4)$$

$$b_n = \frac{m\psi_n(x)\psi'_n(mx) - \psi_n(mx)\psi'_n(x)}{m\xi_n(x)\psi'_n(mx) - \psi_n(mx)\xi'_n(x)} \quad (5)$$

$\psi_n(x)$ *and* $\xi_n(x)$ represent the Riccati-Bessel functions and it is described by the following expression.

$$\psi_n(x) = \sqrt{\frac{\pi x}{2}}\, J_{n+\frac{1}{2}}(x) \quad (6)$$

$$\xi_n(x) = \sqrt{\frac{\pi x}{2}}\, [J_{n+\frac{1}{2}}(x) + i\, Y_{n+\frac{1}{2}}(x)] \quad (7)$$

$J_n(x)$ y $Y_n(x)$ are the Bessel functions of first and second kind respectively [9, 10].

Using the Rayleigh approximation, diffused electromagnetic fields can be approximated, since the first terms of the Mie coefficients are dominant [11, 12].

$$a_1^e = -\frac{i2x^3}{3}\frac{\varepsilon - 1}{\varepsilon + 2} + O(x^5) \quad (8)$$

$$b_1^m = -\frac{i2x^3}{3}\frac{\mu-1}{\mu+2} + O(x^5) \qquad (9)$$

In equation (9), $\mu = 1$ (for materials in the visible range), is important to note in equation (8), when $\varepsilon = -2$, a_1^e becomes infinite, if a_1^e becomes infinite, Q_{ext} and Q_{sca} both become infinite also; this cannot be possible in nature. Nevertheless focusing a particular wavelength on certain metals we can obtain a situation where Q_{abs} and Q_{sca} reached a maximum in determined wavelength, in this case one speaks of Plasmon resonances localized (LSPR), since electron cloud in the metal (electronic plasma) oscillates with maximum amplitude at that wavelength [12]. Although the Mie theory do not occupy of this kinds of plasma oscillations generally the peaks present in the scattering spectra can be interpreted like these characteristic oscillations of plasma [5, 13].

3. Results and Discussion

3.1 Simulations results

The figure 1 shows the simulations of Q_{abs} , Q_{ext} and Q_{sca} *vs* incident wavelength for CuNPs suspended in water, acetone and ethanol, in these results we could observed a particular pattern of peaks corresponding with the oscillation modes of colloidal CuNPs based on the Mie theory. In the simulation we can see the maximum of Mie efficiencies located in the region corresponding to the maximum absorption of radiation for CuNPs. Also, in the simulations we can see a dependence of nanoparticle size and the surrounding medium with peaks positions in the spectra, to change these parameters in the simulations we could observe a shift of the peaks for grater wavelength and widening of the absorption peaks, similarly it happened with NiNPs and CoNPs in the same surrounding media. These simulations were done based on the previous code developed by Mätzler for calculated the Mie scattering [11]. The simulation was done using the values suggested by (Johnson and Christy 1974) [17], for complex refractive index of material and refraction index for medium.

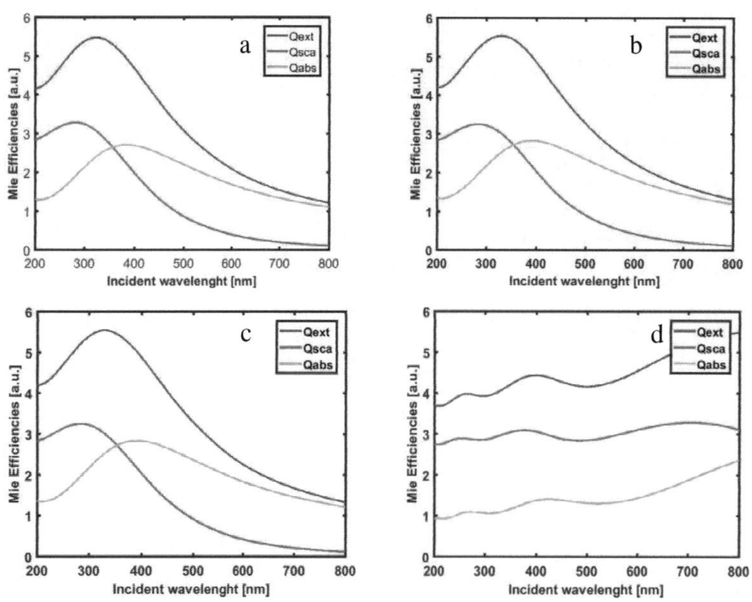

Figure 1: Simulation of Mie efficiencies Q_{abs} , Q_{ext} and Q_{sca} for CuNPs of 40 nm, in different liquid environments a) water n=1,3333, b) acetone n=1,3592 and c) ethanol n=l,3612. In subfigure d) is shown Mie efficiencies for CuNPs in water with 100nm of size.

IOP Conf. Series: Journal of Physics: Conf. Series **850** (2017) 012017 doi:10.1088/1742-6596/850/1/012017

Figure 2 shows electric field intensities vs size parameter, generated by CuNPs suspended in water, the simulations show the radial dependence of electric field intensities emitted for colloidal nanoparticles produced by the surface electron cloud interacting with light. We can see in simulations that electric field presented growth approximately in exponential form with NP's size parameter, and its maximum value is closed to the sphere surface as shown in figure 2, this maximum intensity is due to charge oscillation by the electric field confinement at the sphere surface, due to this increase in the electric field intensities near to surface is that the modes of oscillation in MNPs are generated.

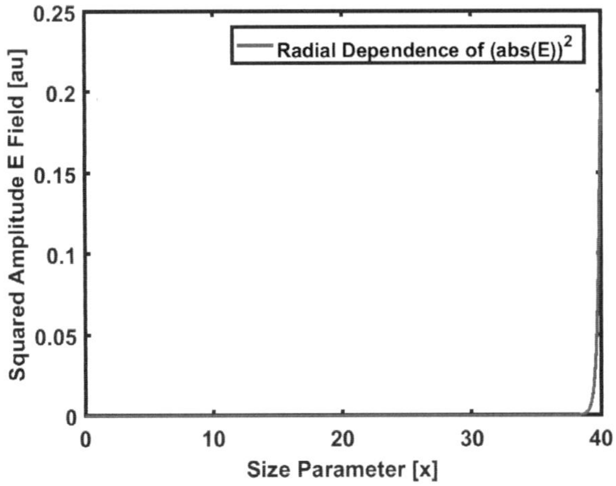

Figure 2: Simulations of radial dependence of electric field intensities in CuNPs suspended in water.

3.2 Experimental details

The laser ablation system to prepared MNPs in this work operates with following parameters: The pressure and temperature remained constants to room temperature. The Nd: YAG laser system (Spectra Physics Indy-10) operated at its fundamental frequency 1064 nm (first harmonic) in Q-Switch mode with rate of 20 Hz, 7 ns of pulse duration and the laser energy to growth was 300mJ. All samples were irradiating during 30 minutes. For the characterization, Evolution 201/220 UV-Visible spectrophotometer from the Thermo Scienti series of quartz cuvettes with 1 cm of optical path was used. Nano-Tracking-Analysis (NTA) was made with the NanoSight NS300 system.

3.3 Experimental results

Figure 3 shows the UV-Vis absorbance spectroscopy produced for CuNPs, NiNPs and CoNPs suspended in deionized water, fabricated by laser ablation method [7]. We observed the principal peaks around 250 nm for CuNPs, 230 nm for NiNPs with most intensity of absorbance due to the ferromagnetic character of Ni. For CoNPs the maximum of absorbance is in 200 nm, and the intensity of absorbance between 400 nm and 700 nm is proximally to cero, it indicated that CoNPs are transparent for the visible radiation. The characteristics peaks of LSPR appear in 590nm, 384 and 350nm like a slight shoulder in the spectrum for CuNPs, NiNPs and CoNPs respectively suspended in water, these results are agreeing with previous results and corresponding with the LSPR band for this kind of MNPs [12, 13]. These results are in agreement with those reported by the literature, which can be corroborated by the similarity of results obtained by H.S Desarkar and collaborators [18], and those obtained in this work. Although the study by Desarkar focuses on the effect of laser radiation time on the sample and its change in absorption spectra, it is important to make this comparison since the same manufacturing method is used and some of the growth parameters are similar.

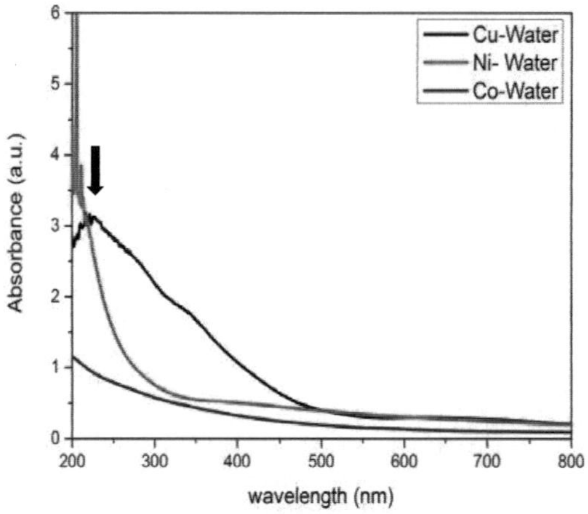

Figure 3: UV-Vis absorbance of colloidal CuNPs, NiNPs and CoNPs suspended in water grown by laser ablation technique.

In figure 4 we could observe the UV-Vis absorbance for CuNPs, NiNPs and CoNPs changing the liquid environment for acetone and we can observe a shifted of the principal peaks, for CuNPs the principal peaks appear in 340 nm and this has widening, which indicates that the CuNPs in acetone are larger than in water, it is important to note that the laser ablation method produces polydispersive NPs solutions. Moreover, that light radiation have a strongly interactions with acetone from 200nm to 300nm region, this is manifested as noise in the acetone NPs spectra. The shift of the maximum to longer wavelengths is a phenomenon of great interest because recently has been used for the design of more efficient solar cells [5]. Also the NPs based on this kind of metals are ferromagnetic at room temperature (ferromagnetic-fluid) and exhibit strong response to the interaction with radiation which as proposed for Chen and collaborators, this phenomenon has been applied to the development of Plasmonics Nano-antennas [15].

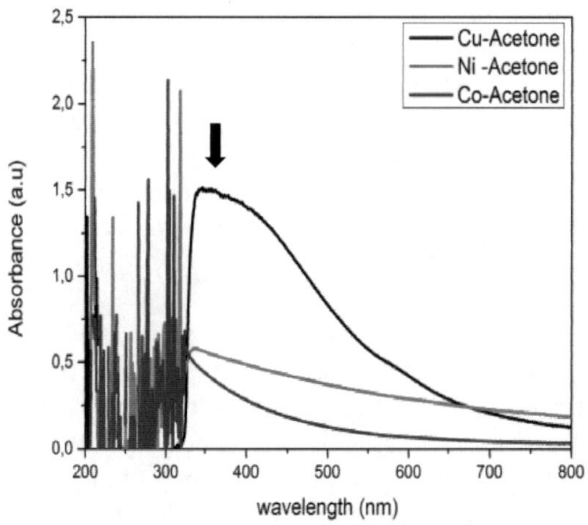

Figure 4: UV-Vis absorbance of colloidal CuNPs, NiNPs and CoNPs suspended in acetone grown by laser ablation technique.

Changing the surrounding media we observed the LSPR most defined for CuNPs suspended in ethanol located at 598nm clearly we can see resonance and we can corroborate a shift of this maximum as seen in the simulation results of the theoretical model and the UV-Vis spectra in figure 5. The peaks shifting were evidenced in the color change of the colloidal dispersions, for example colloidal CuNPs show a green color suspended in water, and colloidal CuNPs suspended in acetone show brown color, another important feature is that CuNPs in water show precipitated and agglomerated with the passage of the hours while CuNPs in acetone did not change their color and did not agglomerate.

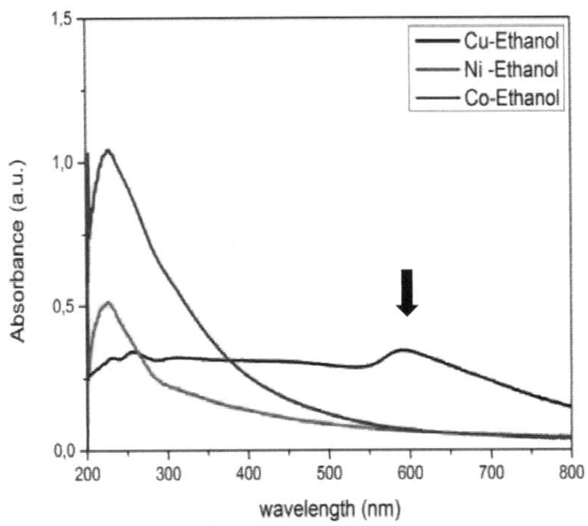

Figure 5: UV-Vis absorbance of colloidal CuNPs, NiNPs and CoNPs suspended in ethanol grown by laser ablation technique.

With this result we can observe that LSPR not only depend on the size and shape of the nanoparticles. If not also the polarization of the medium, for a higher index of refraction the dielectric function increases and therefore there will be greater polarization of the medium attenuating the accumulated charge in the resonance zone. By attenuating the electric charge of the NPs this reduces the restoring electric force, and reducing the restoring force immediately reduces the resonance frequency, that is why the absorption peaks present red-shifted.

Figure 6 shows Nano Tracking Analyzed (NTA) results in this results we can see the light scattering produced by CuNPs, NiNPs and CoNPs. In addition in figure 6a, we can see the nanoparticle-light interaction shown the characteristic scattering pattern of CuNPs, we can see the increased intensity of electric field near the edge of the nanoparticle and its strong interaction with other nanoparticles forming a characteristic dipole scattering pattern. We could found that CuNPs and NiNPs show greater absorption of radiation it is evident in NTA results show in figure 6b exhibit the intensity of CuNPs suspended in water and his relation with size of the particles, it is according with our simulations results showed in the figure 2.

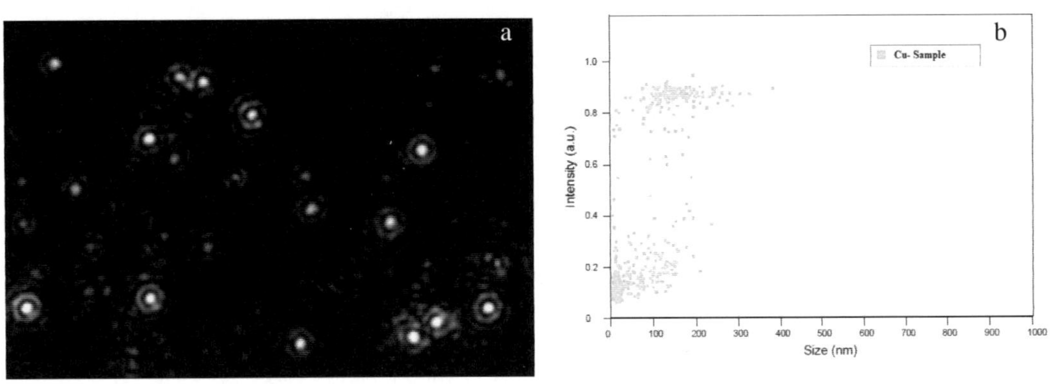

Figure 6: a) Scattering electric field pattern of MNP's using NTA method. b) Distribution of electric field intensities produced by CuNPs of different size suspended in water, grown by laser ablation technique.

4. Conclusions

In summary, a theoretical model to described the light scattering by MNPs using Mie theory were studied and simulated, also we could found the LSPR in CuNPs, NiNPs and CoNPs and determine its depending whit the size and surrounding medium, these resonances were due to charge oscillation close the sphere surface and there show the maximum intensity of electric field, also we can see the electric field grow approximately in exponential form with NP's radius just concentrate on the surface of the sphere. We observed the principal peaks at 250 nm, 230nm and 200nm for CuNPs, NiNPs and CoNPs respectively corresponding with interband transitions. The Localized plasmon resonances appear in 590nm, 384 and 350nm for CuNPs, NiNPs and CoNPs suspended in water. LSPR most defined was for CuNPs in ethanol, and in 598nm clearly we could see these resonances by UV-Vis spectroscopy, also we could determine that the colloid of CuNPs in acetone were the most stable solution.

Acknowledgments
This work was supported by a grant from the *Convocatoria 738 jovenes investigadores alianza COLCIENCIAS-SENA 2015* (Colombia).

References

[1] Gustav Mie. 1908. Ann. Physics, 25(4):377-445.
[2] Bohren, C. and D. Huffman, 1983. John Wiley and Sons, New York.
[3] Xiaofeng Fan, Weitao Zheng, and David J Singh. 2014. Light: Science and Applications, 3(79): 1-14,
[4] Daniel A. Cruz and Miriam C. Rodríguez and Juan M. López, (2012). Avances en Ciencias e Ingeniería 2-3, 67-78.
[5] Hossein Ghaforyan, Majid E., and Sara Mohammadi. 2015. TI Journals World Appl. Programming, 5(4):79-82.
[6] Charles Pool and Frank Owens. Introduction to nanotechnology. 2003. John Wiley and Sons, Inc. 107-110.
[7] J.S.Duque, H. Riascos, G. Osorio and J.P Cuenca 2015. El hombre y la máquina. E 47. 37-42.
[8] Guillaume Baffou. 2012. Mie theory for metal nanoparticles. CNRS 1-2.
[9] John David Jackson. 1999. John Wiley and Sons, Inc. Third Edition.
[10] Kou-Nan Liuo. 1977. University Utah. University Utah.
[11] Christian Mätzler. 2002 . Research Report No. 200208, Institut für Angewandte Physik.
[12] Daniel Mazón Solórzano. 2012. Proyecto fin de Carrera Universidad de Cantabria.
[13] V.A.G. Rivera, F.A. Ferri and E. Marega Jr. (2012).Plasmonics - Principles and Applications, Dr. Ki Young Kim (Ed.), InTech, DOI: 10.5772/50753.
[14] M. Muniz-Miranda · C. Gellini · A. Simonelli · M. Tiberi · F. Giammanco · E. Giorgetti, 2013. Appl Phys A 110:829–833 DOI 10.1007/s00339-012-7160-7.
[15] Jianing Chen , Pablo Albella , Zhaleh Pirzadeh , Pablo Alonso-González ,Florian Huth, Stefano Bonetti (2011). Small 2011, *7,* No. 16, 2341–2347. DOI: 10.1002/smll.201100640.
[16] Helmuth Horvath. 2009. Journal of Quantitative Spectroscopy & Radiative Transfer (110) 787–799.
[17] P. B. Johnson and R. W. Christy. 1972. Phys. Rev. B 6, 4370.
[18] H. S. Desarkar P. Kumbhakar A. K. Mitra. 2012. Appl Nanosci 2:285291, pags. 285-291.

IOP Conf. Series: Journal of Physics: Conf. Series **850** (2017) 012018 doi:10.1088/1742-6596/850/1/012018

Seismic migration in generalized coordinates

C. Arias, L. F. Duque

Centro de Ciencias de la Computación, Instituto Tecnológico Metropolitano, Medellín, Colombia

E-mail: lduquego@gmail.com

Abstract. Reverse time migration (RTM) is a technique widely used nowadays to obtain images of the earth's sub-surface, using artificially produced seismic waves. This technique has been developed for zones with flat surface and when applied to zones with rugged topography some corrections must be introduced in order to adapt it. This can produce defects in the final image called artifacts. We introduce a simple mathematical map that transforms a scenario with rugged topography into a flat one. The three steps of the RTM can be applied in a way similar to the conventional ones just by changing the Laplacian in the acoustic wave equation for a generalized one. We present a test of this technique using the *Canadian foothills* SEG velocity model.

1. Introduction

RTM is an imaging technique that although was introduced in the year 1983 [1], it has only began to be used in the recent years because the computational resources needed to implement it were commonly available recently. Despite of its high computational cost, RTM is nowadays the better choice among a big set of options to produce seismic images because it can be used in zones with strong variations on the velocity of propagation; it can map sub-surface structures with any dipping angle (unlike other techniques like those based on one way wave equations (OWWE) [2, 3, 4]) and can create good images of zones of interest, like those under and around salt domes where petroleum reservoirs can be found.

The RTM method has been developed for zones with a flat acquisition surface, i.e., zones where the controlled seismic waves are generated and registered in the geophones should be flat. The application of this method to zones with strong variations on topography requires the forced application of a cartesian mesh to a curved domain and this can lead to a wrong ubication of the sub-surface structures as the curved surfaces should be modeled in a ladder way with steps of limited size. Since the acquisition grid has a limited size the ladder effect is not easy to avoid and furthermore if smaller step sizes are introduced to model in a better way the curved boundary a increased computational cost would be obtained.

A RTM algorithm for domains with curved boundary has been proposed recently [5]. This approach was based on a complex variable transformation for 2D domains. A more general approach was introduced by Shragge in 2014 [6] in which the 3D acoustic wave equation was solved for 3D domains. The next step should be to implement the complete RTM algorithm, reason why this work was developed here. We present a simpler map that transforms a generally curved acquisition surface into a flat one. The curved domain is transformed into a rectangular domain where a uniform grid can be applied to solve the acoustic wave equation by using the generalized Laplacian. When the 3 steps of the RTM are finished in this rectangular domain, the final image can be mapped to the curved domain. In this way this method can create images for reflectors under zones like mountains of foothills.

2. Method

The transformation used for mapping a rectangular domain with coordinates (ξ_1, ξ_2) (named computational domain) into the physical domain of coordinates (x_1, x_2) is

$$x_1 = \xi_1 \tag{1}$$
$$x_2 = \xi_2 + \phi(\xi_1), \tag{2}$$

where $\phi(x_1) = \phi(\xi_1)$ is a smoothed function that represents the curved upper boundary of the physical domain. This transformation is depicted in Figure 1.

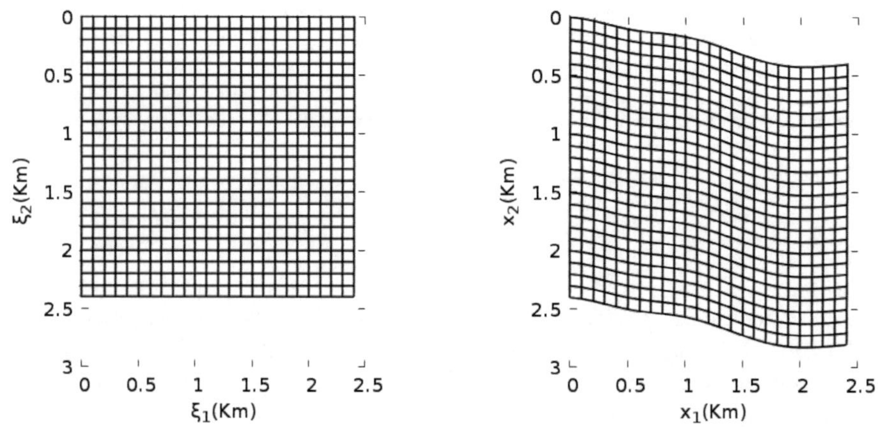

(a) (b)

Figure 1. Transformation from the computational domain (a) to the physical domain (b)

Using change of variables from (1) and (2) we find the new expression for the Laplacian operator, by using chain rule or we can use the standard form in generalized coordinates. The acoustic wave equation for the transformed domain is given by

$$\frac{\partial^2 P}{\partial t^2} = c^2(\xi_1, \xi_2)\nabla^2 P + f \tag{3}$$

where

$$\nabla^2 = \frac{1}{\sqrt{|g|}}\frac{\partial}{\partial \xi_i}\left(g^{ij}\sqrt{|g|}\frac{\partial}{\partial \xi_j}\right), \qquad i, j = 1, 2 \tag{4}$$

and f is the source term.

Equation (4) gives the generalized Laplacian, where $|g|$ is the absolute value of the determinant of the metric tensor g_{ij}, given by

$$g_{ij} = \frac{\partial x_k}{\partial \xi_i}\frac{\partial x_k}{\partial x_j}. \tag{5}$$

In order to re-write the wave equation, a contravariant representation of the metric tensor $g^{ij} = g_{ij}^{-1}$ [7] is needed, where the sum over repeated indexes is implied. Note that c^2, the square of the velocity vector, is a scalar and therefore it is not transformed. However, its arguments are transformed. Expanding the Laplacian we can re-write Equation (4) in a more convenient way as:

$$\nabla^2 = \zeta^j \frac{\partial}{\partial \xi_i} + g^{ij} \frac{\partial^2}{\partial \xi_i \partial \xi_j} \tag{6}$$

where

$$\zeta^j = \frac{1}{\sqrt{|g|}} \frac{\partial}{\partial \xi_i} \left(\sqrt{|g|} \right). \tag{7}$$

The elements ζ^i are geometric factors so do are $g's$, so they have to be calculated only once. For our specific transformation, given in expressions (1) and (2), the equations (5) and (7) give

$$g^{ij} = \begin{pmatrix} 1 & -\frac{\partial \phi}{\partial \xi_1} \\ -\frac{\partial \phi}{\partial \xi_1} & 1 + \left(\frac{\partial \phi}{\partial \xi_1}\right)^2 \end{pmatrix}, \tag{8}$$

$$\zeta_1 = 0, \tag{9}$$

$$\zeta_2 = \frac{d^2 \phi}{d\xi_1^2}, \tag{10}$$

$$\det g = 1. \tag{11}$$

Using expression (6) for the Laplacian, along with (8), (9) and (10) equation, the wave equation (3) can be solved in the computational domain. The forward and backward propagation and imaging condition can all be done in the regular grid. To obtain the RTM image $I(x_1, x_2)$ the standard cross correlation between the forward P_f and the backward P_b propagated fields can be used:

$$I(\xi_1, \xi_2) = \sum_s \sum_r \int dt P_f(\xi_1, \xi_2, t) P_b(\xi_1, \xi_2, t) \tag{12}$$

where the first sum is over all the receptors, the second over all sources and the second is over all receptors and the integral is over time. The image $I(x_1, x_2)$ can be obtained form $I(\xi_1, \xi_2)$ just by transforming the arguments from (ξ_1, ξ_2) to (x_1, x_2) using the transformation equations (1) and (2).

The stability condition for this method can be derived in an heuristic way [5]: the standard Courant condition is

$$\Delta t \leq \frac{\Delta r}{v_{\mathbf{x}}}, \tag{13}$$

where

$$\Delta r = [\Delta x_1^{-2} + \Delta x_2^{-2}]^{-1/2}, \qquad \Delta x_i \approx \frac{\partial x_j}{\partial \xi_i} \Delta \xi_j, \tag{14}$$

so equation (13) gives

$$\Delta t \leq \frac{1}{v_\xi} \text{arg min} \left[\left(\frac{\partial x_1}{\partial \xi^j} \Delta \xi_j \right)^{-2} + \left(\frac{\partial x_2}{\partial \xi^j} \Delta \xi_j \right)^{-2} \right]^{-\frac{1}{2}} \tag{15}$$

3. Results

We applied the method described above to create an RTM image for the Canadian Foothills velocity model, which is a synthetic velocity model for a region in British Columbia that shows several complex geological structures and a rugged topography, common in that Canadian zone [8]. The synthetic seismograms were generated solving the wave equation in the computational domain and taking the field at $\xi_2 = 0$, which takes into account the transformation equation (2) corresponding to $x_2 = \phi(x_1)$.It means points over the mountain border since the geophones are supposed to be placed in the mountain line. The original Canadian foothills model has size 1668x1000 but we used a sub-sampled version of size 300x200 (taking 1 sample each 5 points in the vertical direction and in the horizontal direction) shown in Figure 2. We took $\Delta x_1 = 0.075Km$, $\Delta x_2 = 0.05Km$, $\Delta t = 0.001s$ and the frequency of the Ricker wavelet used for the source term was 7 Hz.

The velocity model should be transformed tho the computational domain. The result is shown in Figure 2. The RTM image is shown in Figure 3.

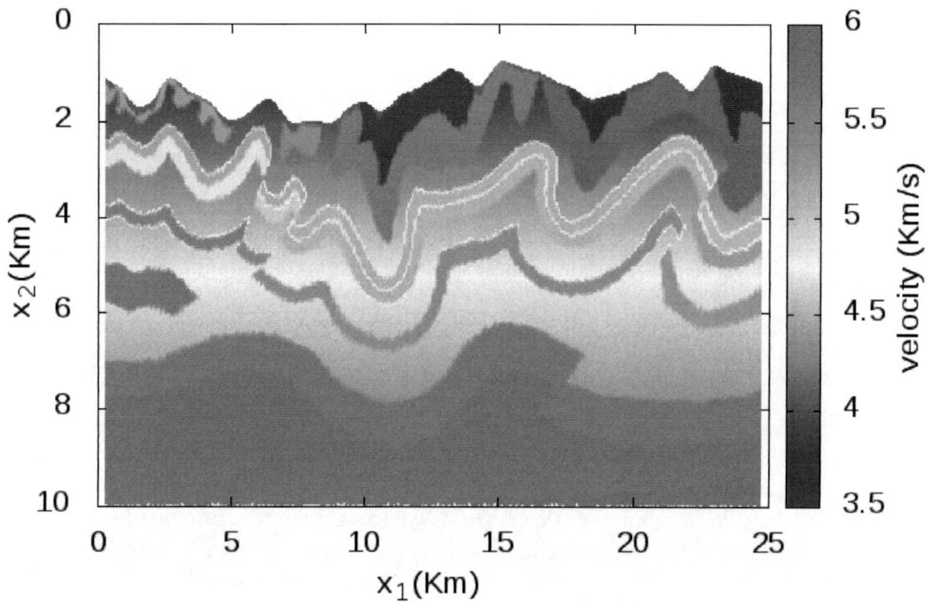

Figure 2. sub-sampled Canadian foothills model.

4. Conclusions

After modifying the Laplacian, by using the transformation presented, the RTM algorithm can be implemented as a classic RTM in a rectangular mesh. The algorithm can handle strong variations, both in topography and in the velocity.

The same transformation can be applied to other problems like the elastic wave equation, for FWI and RTM algorithms. In that case the approach can be even simpler: changing the derivatives in the physical domain for derivatives in the computational domain using the chain rule.

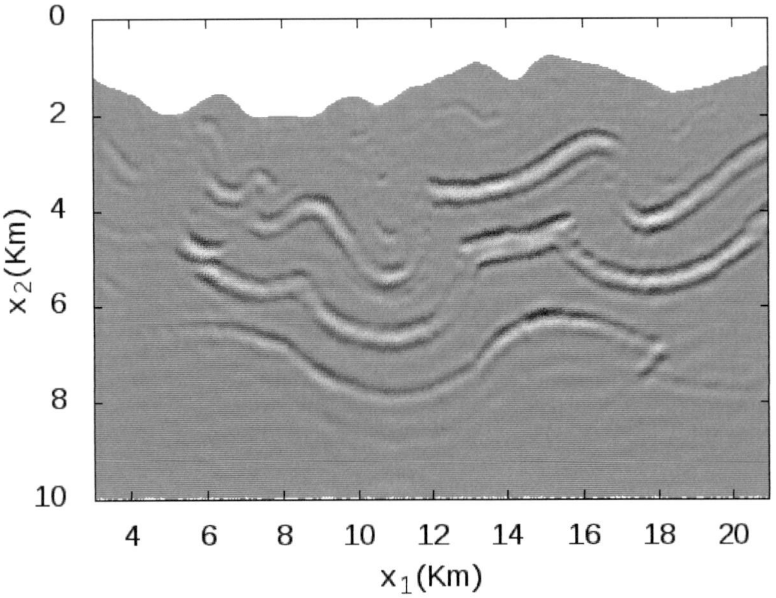

Figure 3. Migrated image

References

[1] Baysal E, Kosloff D D and Sherwood J W C 1983 Reverse time migration *Geophysics* **48** 1514
[2] Claerbout J and Doherty S 1972 Downward continuation of moveout corrected Seismograms *Geophysics* **28** 60
[3] Loewenthal D, Lu L, Roberson R and Sherwood J W C 1976 The wave equation applied to migration *Geophys. Prosp.* **24** 380
[4] Stolt R 1978 Migration by Fourier transform *Geophysics* **43** 23
[5] Shragge J, 2014 Reverse time migration from topography *Geophysics* **79** No 4 1
[6] Shragge J, 2014 Solving the 3D acoustic wave equation on generalized structured meshes: A finite-difference time-domain approach *Geophysics* **79** No 6 1
[7] Synge J L and Schild A 1978 *Tensor calculus* (Mineola, NY:Dover Publications)
[8] `http://software.seg.org/datasets/2D/Model_1994/`

IOP Conf. Series: Journal of Physics: Conf. Series **850** (2017) 012019 doi:10.1088/1742-6596/850/1/012019

Simple-1: Development stage of the data transmission system for a solid propellant mid-power rocket model

Andrés Yarce[1,2] ,Juan Sebastián Rodríguez[1], Julián Galvez [3], Alejandro Gómez[2], Manuel J. García[1,4]

[1] Applied Mechanics Research Group, Engineering School, Universidad EAFIT
[2] Mathematical Modeling Research Group GRIMMAT, School of Sciences, Universidad EAFIT
[3] Computational Physics and Astrophysics Research Group Facom and Applied Telecommunications Research Group GITA, Universidad de Antioquia
[4] Mechanical Engineering Department,University of Texas at San Antonio

E-mail: ayarceb@eafit.edu.co, jrodri36@eafit.edu.co, julian.galvezs@udea.edu.co, agomez13@eafit.edu.co, manuel.garciaruiz@utsa.edu

Abstract. This paper presents the development stage of a communication module for a solid propellant mid-power rocket model. The communication module was named. Simple-1 and this work considers its design, construction and testing. A rocket model Estes Ventris Series Pro II® was modified to introduce, on the top of the payload, several sensors in a CanSat form factor. The Printed Circuit Board (PCB) was designed and fabricated from Commercial Off The Shelf (COTS) components and assembled in a cylindrical rack structure similar to this small format satellite concept. The sensors data was processed using one Arduino Mini and transmitted using a radio module to a Software Defined Radio (SDR) HackRF based platform on the ground station. The Simple-1 was tested using a drone in successive releases, reaching altitudes from 200 to 300 meters. Different kind of data, in terms of altitude, position, atmospheric pressure and vehicle temperature were successfully measured, making possible the progress to a next stage of launching and analysis.

1. Introduction

Colombia has an incipient aerospace sector, with only one satellite in orbit named Libertad 1, designed by Universidad Sergio Arboleda between 2004 and 2007 for academic purposes [1]. In the same way, at industrial level, the situation is not very different, as there are few companies working on space exploration, there is not yet a critical mass of initiatives to enter into the demanding aerospace industry. Mostly advances in aerospace topics are in academy, which is the case of Proyecto Uniandino Aeroespacial (PUA) planned by Universidad de los Andes [2] and most recently the Universidad de Antioquia with the undergraduate program in Aerospace Engineering [3].

There is a gap in the aerospace thematics in Latin America [4]. Space exploration has allowed the creation of new market prospects and new technologies that have improved our global economy and have achieved improvements in our lifestyle [5]. A way to close the gap in countries without previous research on these issues is throughout CanSat missions [6] and the experimentation with rockets projects, as an useful strategy to educate people by training in hands-on experience activities. Training of people will also help in the development of multidisciplinary skills useful in more complex missions such as project management,teamwork abilities, manufacturing process, design of experiments, communications subsystem development and big data analysis [7, 8, 9].

CanSat is a standard form factor of a nano-satellite with a volume around 350 ml and a mass about 500 g, that has a shape of a cilindrical rack structure [10, 11, 12]. All basic functions of a satellite are represented in CanSats, such as the Energy Power Supply Subsystem (EPS), the Communications Subsystems, Command On Board & Data Handling (CO&DH) and Payload Subsystem. A CanSat could perform several experiments such as altitude control measurements, image capture, downlink of several sensors [13, 14, 15], and differential GPS measurement.

Electronic applications such as flight instrumentation, avionics or communications subsystems can be tested in sounding rockets or even in some commercial rocket models, because they provide a practical form to test extreme condition such as thrust, vibrations, acceleration, atmospheric pressure and thermal variations.

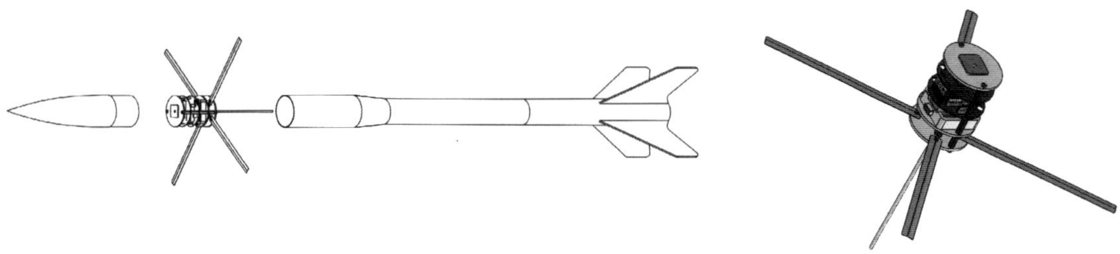

Figure 1: Simple-1 CAD model distribution detail

The Simple-1 mission was created as an initial step to contribute in the development of the aerospace topics in our universities and region. The objective of this mission was to learn about the manufacturing processes of components for the emergent aerospace industry in our country.The total integration scheme of the mission is shown in Figure 1. The name of the mission was not selected by a deliberate consensus, it was just a curious way of referring to the mission, in an attempt to appropriate design methods that reference the KISS principle (*Keep It Super Simple* [16]), by means of selecting items that using low power transmission, low consumption and easy to get.

In the first part of the paper the critical aspects of electronic design of the mission, and the technical specification of the components used are documented. Secondly the results are exposed. Finally the last part of the paper the conclusions, the discussions and future work recommendations for the following stages of testing and launching of the Simple-1 mission are given.

2. Methodology

Dynamic and structural factors of the rocket were considered to make the selection of the components according to the characteristics of useful weight and its trajectory. The Simple-1 is divided in four subsystems: The communication, the Energy Power Supply (EPS), the Command on Board & Data Handling (CB&DH) and the payload subsystem. For the communication subsystem the radio module were selected and the antenna were manufactured. The EPS was designed considering the power requirements of all the components and the operation time of the mission. The CB&DH was selected taking into account the microprocessor capabilities. When the basic subsystems were established, the payload, consisting of several sensors, was designed and integrated.

2.1. Rocket dynamic conditions

Once the commercial rocket model and their payload weight capacities were chosen according to the volume available, the geometry of the rocket was subjected to dynamic simulation using open source OpenRocket Software, as shown in Figure 2. This software uses the Runge-Kutta 4 (RK4) integration method as a numerical solver, which is a fourth-order integration method; for this reason, the total simulation error is of the order $0(\Delta t^4)$.This simulation was made with the goal to determine the maximum velocity and acceleration factors in the trajectory estimated for the solid propellant motor [17, 18].

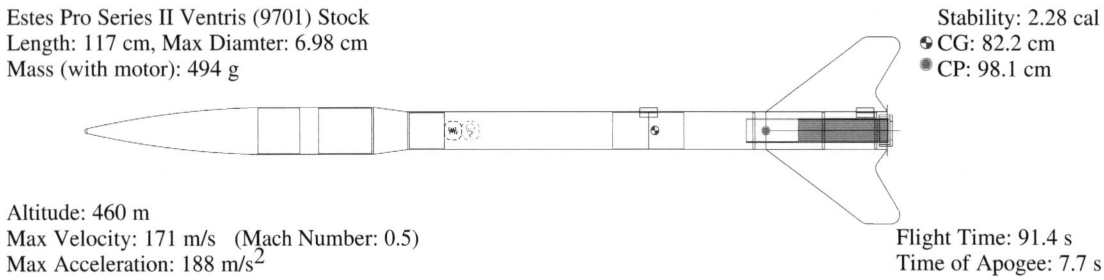

Figure 2: OpenRocket simulation view at initial position and orientation the rocket at time *t=0*.

This rocket has a dead weight equal to 500 g and reaches an approximate altitude of 460 m according to simulations using a Estes composite rocket motor (G80-7T) that generates a maximum thrust of 108 N.

2.2. Electronic components selection

According to the design requirements for the mission the electronic components were selected. Some characteristics of the Simple-1 electronic module are shown in Table 1.

2.3. Communication system

The communication subsystem was divided in two groups: The flight segment and the ground segment. The flight segment was composed by the Radiometrix (NTX2-434.650) module, several SMD (Superficial Mounting Device) LED indicators, a SMA (Surface Mounting Antenna) connector and a coaxial antenna cable adapted to the antenna. The antenna consist of quadrupole geometry that works as a ground plane and the antenna radiator which was calculated for the wavelength of the radio-frequency used (16.4 cm). The sensors data is send to the Radiometrix from the Arduino using a PWM signal that assigns a specific voltage level shifting between two values that represent different frequencies from the antenna. The Figure 3 shows the operation schematic of the subsystems on flight.

The second segment is a ground station based on a Software Defined Radio SDR architecture. The radio used was a HackRF. The data was received via a link between the software to receive data from SDR radios named SDR# or GQRX with the decode software DL-FLDIGI throughout a virtual audio cable. The Figure 4 shows a diagram of the operation of the communication ground station subsystem.

Table 1: Electronic Components: Simple-1 Module

Component	Reference	Features
Main onboard computer	Arduino Mini Pro	• Clock: 16 MHz • Flash Memory: 32 KB • SRAM: 2 KB • Microcontroller: ATmega328
Data Storage	MicroSD Data Logger	• Flash Memory: 8 GB
Data Transmitter	Radio Metrix NTX2	• Frequency: 434.650 MHz • Channel Spacing: 25 kHz • Data Rate: 10 kHz
Position Measurement	GPS Ublox NEO 6M GY-GPS6MV2	• Update Rate: 5 Hz • Position Accuracy: 2.5 m • Velocity Accuracy: 0.1 m/s
Inertial Measurement	3Ch Accelerometer LSM303	• $\pm2, \pm4, \pm6, \pm8, \pm16g$ at 16bits
Unit IMU	3Ch Magnetometer LSM303	• $\pm2, \pm4, \pm8, \pm12G$ at 16bits
	3Ch Gyroscope L3GD20	• $\pm245, \pm500, \pm2000°/s$ at 16bits
	Barometric Pressure BMP180	• $+9-0.5$ at 16bits
	Temperature Sensor BMP180	• $0to65°C$ at 16bits
Energy Storage	Battery LiPO	• $7.2V$, $200mA$

2.4. Energy Power Supply subsystem (EPS)

The EPS subsystem is composed by two LiPo batteries joined in series, a mini DC-DC Buck Step Down power conversion module, a switch and a capacitive element to reduce the noise rate level of the power input signal to feed the Arduino. A connector to charge the battery and to access the payload inside the rocket structure is part of this subsystem. A structure to hold the batteries in place was manufactured using a 3D printer.

2.5. On Board Data Handling (OB&DH)

The OB&DH subsystem of the Simple-1 was composed by the Arduino pro mini (Atmega168) and a data external storage using a MicroSD Datalogger. The Arduino module receives all the data from the sensors connected and distribute its capacity by saving data into the MicroSD datalogger and sending packets to the Radiometrix module. These two process have different velocities, saving more data per unit of time that the data per unit of time of the Radiometrix channel. The total code size is:

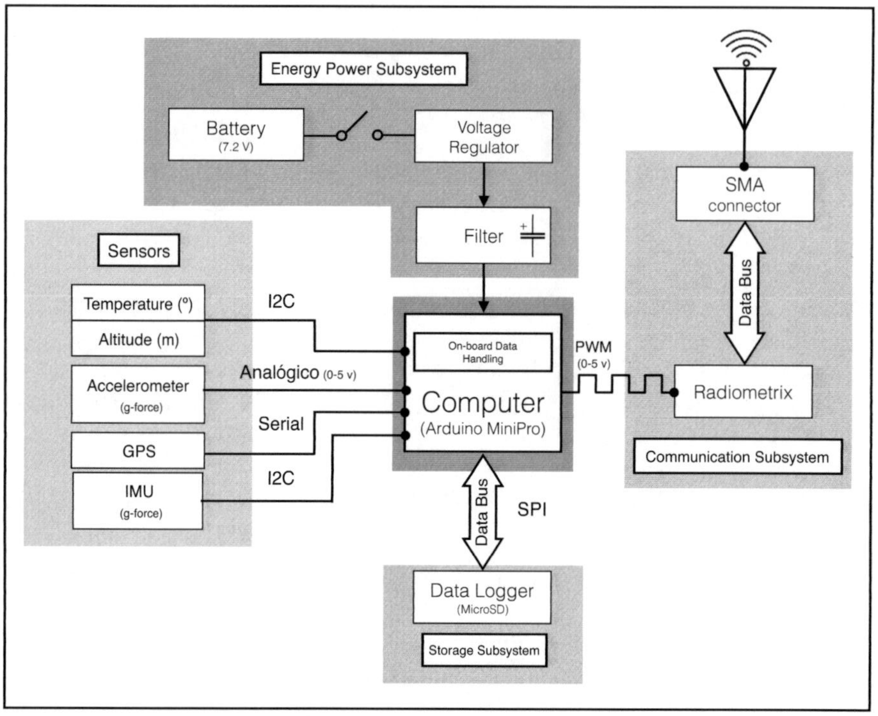

Figure 3: Flowchart of the subsystems on flight, Simple-1 module

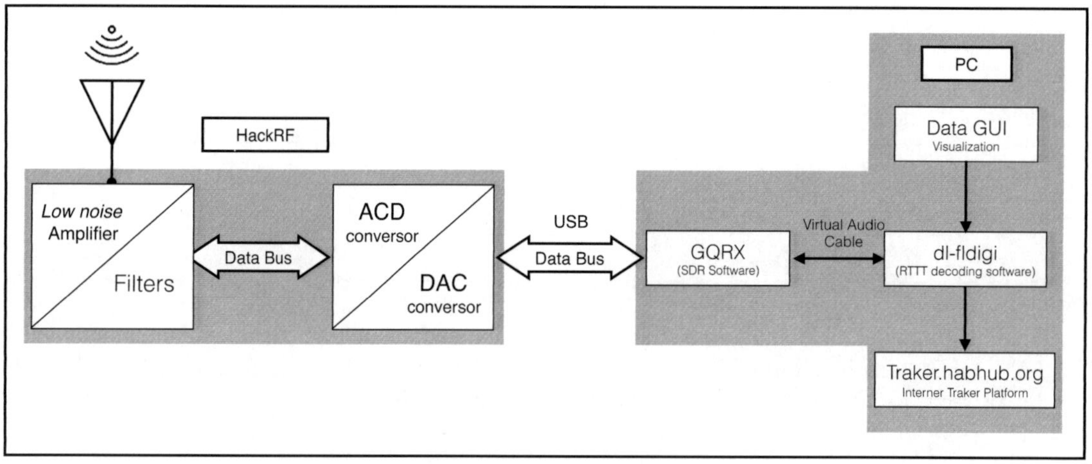

Figure 4: Flowchart SDR ground station

- Program memoria used: 27.158 bytes (88%)
- Dynamic memory used: 1.504 bytes (73%)

The use of the 8-bit micro-controller limited the amount of computing capabilities and the libraries were reduced to optimize it. Transmission of information through the output of the Radiometrix at 50 Bd was successfully achieved.

2.6. Payload (Sensors)

Generally, the Cansat missions as rocket's payload, besides of having internal storage subsystems, are designed to send in the more near real-time possible data transmission from the remote measurements devices in the descent flight. The intention was to test the concept of using a communications subsystem that provides information of different stages of the rocket flight, considering the ascent and descent time and the redundancy that the storage subsystem (SD) provides. The sensors of the Simple-1 compose the Payload subsystem: the analog accelerometer (x, y, z), the Inertial Measurement Unit (IMU), the GPS and the temperature and altitude sensor.

3. Results

3.1. Altitude Simulation

According to the simulation of the rocket with its characteristics and the payload (Figure 5), it was estimated that the flight time is around 91.4 s, reaching a maximum altitude of 460 m. That

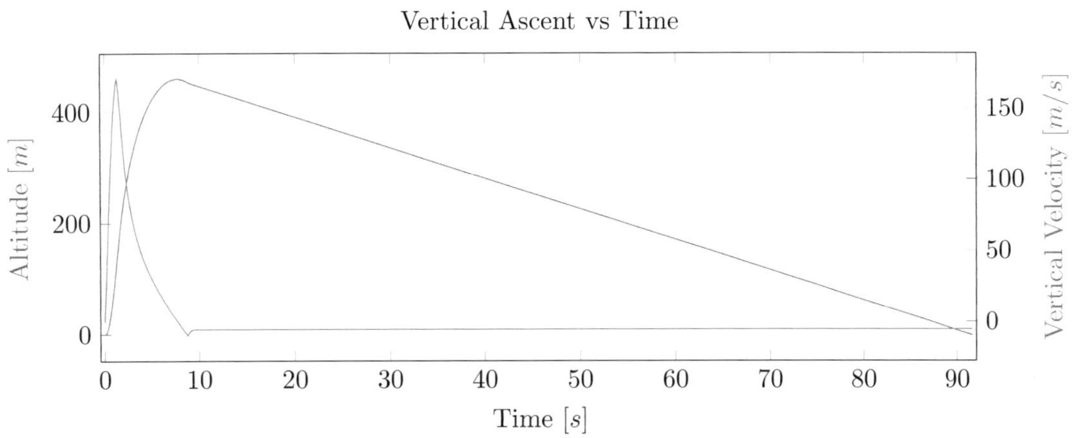

Figure 5: Plot of the flight simulation trajectory Estes Ventris Series Pro II®

3.2. Payload

The payload was designed from the dimension considerations of the Estes rocket in order to introduce it in the upper internal part. The radiator monopole was assembled in the center of the array ground plane structure. The dipoles of the ground plane were placed in the orthogonal plane to the axis of the rocket what makes them excel from the structure of it. Due to the fact that the antenna ground plane dipoles are attached to the rocket structure, it was necessary to accomplished a Drone altitude test to check the range and stability of the signal transmitted. The chemical etching method was used to generate the tracks of the circuit as a practice to generate some experience in the team about the PCB manufacturing process. First the board was designed using Eagle and then was printed and transferred over the copper with heat, the etching method was implemented. (Figure 6).

This first approach of the module allowed a successful transmission of data from sensors attached on it. This allows the formalization of the electronic design methods, and also allow reduce the cost and time for the next printed circuit using an industrial process. The final board is presented in Figure 7.

IOP Conf. Series: Journal of Physics: Conf. Series **850** (2017) 012019 doi:10.1088/1742-6596/850/1/012019

Figure 6: First PCB elaborated.

Figure 7: Final PCB.

3.3. Drone altitude test

Quadricopter tests were made in order to prove the transmission distance and calibrate the payload. The Simple-1 was elevated to altitudes in the range between 200 to 300 m as is shown in the Figure 8.

Figure 8: Simple-1 being tested in a quadricopter

In the first test some noise was detected in the data packets but insignificant. The first experiences were made transmitting at a data rate of 50 Bd.

3.4. Distance test

A long distance test has been implemented, seeking to measure the capacity of the transmission module. In this test two static positions were established, the T position corresponds to point data transmission, and the R position corresponds to the point of receiving the data. Both points will have a range among themselves of 1.3 Km in the city of Medellín. The geographical location of these points is in the Figure 9. This test allowed to verify the distance of transmission of the payload. The Data packets were transmitted at a rate of 300 Bd, with a maximum gain of 43 dB and a minimum gain of 24 dB. The importance of this test is that was conducted in a real environment, with natural and artificial obstacles, which affect the transmission of data. It should be noticed that this type of environment, in terms of radio frequency data transmission, can become an hostile environment for the amount of noise and electromagnetic interference.

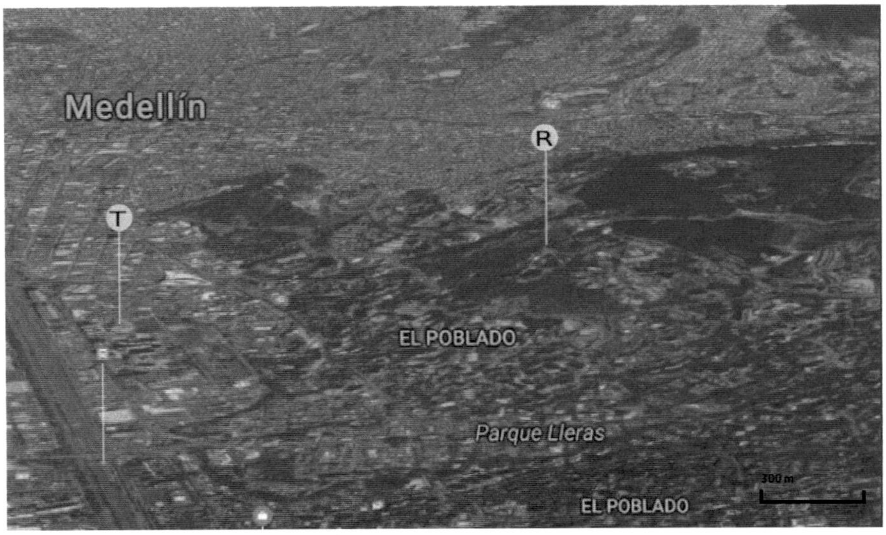

Figure 9: Geographical location of the long distance test point

4. Conclusions

In order to reduce the volume of this kind of modules is necessary to get experience in the electronic development using Surface Mounting Devices (SMD). The test communication procedures of the CanSat Simple-1 made in the drone flight and in the test between two sight line points allowed to transmit data successfully and check the transmission range of the payload. The system should attempt to shield the circuits in order to prevent different sources of noise. This is a detailed activity that requires knowledge to characterize individually each components to generate the appropriate ground shields for the components that required it and to make the appropriate physical distribution of the components and subsystems for next developments. Is important to understand the frequency noise that generate each subsystem and also understand more about external radiation source to prevent the integrity of the module.

Discussions

In future work it is recommend the use of two independent computing modules for the communication, the data reading and storage subsystem because each one needs a critical computing time operation. The use of at least two modules of each sensor or system on board is a good practice to improve the exactness of the measurements.

Is important to improve the fact that the communication module is sensitive to temperature changes increasing or decreasing . A shift in peak frequency is noticed with the temperature variation and this make difficult the signal tracking. Was necessary to track the signal due to the change to keep the transmission stable. Is important to think in temperature stabilization mechanisms for this module in the next future developments.

The pseudo-analog signal sent to the Radiometrix using the PWM of the Arduino probably has big influence to the noise of the signal transmitted because it has a frequency signal of 36 kHz that was chosen over de default low PWM frequency that present an poor performance in the ration signal-noise. Is recommended to ensure a cleaner and stable signal without sacrificing response time between high-voltage and low bit because it could disable communication transmissions, using low pass filters according to the frequency of the RTTY data speed used.

The radio that was used is more commonly suitable for high altitude balloon missions because the data rate transmission are less demanding. It is important to assume considerations like the data rate transmission before the design of the communication subsystem. If the interest is in the measures during the ascent stage and also transmission is important to think in a radio with more fast data rate transmission, if the focus on the mission is to collect data and not necessary the near real-time transmission the descent period will be enough to collect and transmit some data.

Acknowledgments

The project was funded by the Research Direction at Universidad EAFIT through the Formative Research Program and, simultaneously, with the efforts and resources of the groups involved. Also, this project is based on the cooperative work between Rocket and Propulsion Study Research Group from the EAFIT's School of Engineering, and the Theoretical and Computational Physics Study Research Group from the EAFIT's School of Science. The authors would like to thank Professor Francisco Javier Botero of Applied Mechanics Research Group for believe in the relevance of aerospace initiatives. Acknowledgements to Mauricio Fernández of the Design Engineering Research Group - GRID for his assistance in the design of quadcopter's test. Finally, the authors thank to the Manufacturing Innovation Study Research Group of the Production Engineering Department for the 3D printer parts sponsorship.

References

[1] Portilla J G 2012 *Revista Académica Colombiana de Ciencias* **36** 491–500 ISSN 0370-3908
[2] Faculty of Engineering U d l A Pua: Proyecto uniandino aeroespacial pua URL `https://pua.uniandes.edu.co/doku.php`
[3] Cañas E Nuevo programa de ingeniera aeroespacial
[4] Sarli B, Cabero M, Lopez A, Cardoso J, Jimenez D, Roman-Gonzalez A, Villena G, Vargas-Cuentas N I and Perazzo F 2015 *66th International Astronautical Congress-IAC 2015* p 11
[5] Lockney D, Schwerin B and Rademakers L 2010 National Aeronautics and Space Administration: Spinn Off document 2010 Tech. rep.
[6] Bautista-Linares E, Morales-Gonzales E A, Herrera-Cortez M, Narvaez-Martinez E A and Martinez-Castillo J 2015 1–4
[7] Nakasuka S, Sako N, Sahara H, Nakamura Y, Eishima T and Komatsu M 2010 *Acta Astronautica* **66** 1099–1105 ISSN 00945765 URL `http://dx.doi.org/10.1016/j.actaastro.2009.09.029`
[8] Yamaura S, Akiyama H and Kawashima R 2011 *RAST 2011 - Proceedings of 5th International Conference on Recent Advances in Space Technologies* 856–860
[9] Miyazaki Y and Yamazaki M 2013 *RAST 2013 - Proceedings of 6th International Conference on Recent Advances in Space Technologies* 1081–1086
[10] Soyer S 2011 *RAST 2011 - Proceedings of 5th International Conference on Recent Advances in Space Technologies* 789–793
[11] Gansmoe T, Mathisen S V, Grande J, Nielsen J F D and Rossing N K 2013 *The CanSat Book*
[12] Ykis H A, Apacak R Y, Agirbas O, Abur S and Soyer S 794–799
[13] Nehme P H D, Borges R A, Cappelletti C and Battistini S 2014 1–7 ISSN 1095-323X
[14] Gopal G, Harith B, Raj R, Savyasachi J and Umadi C 2016 **6** 5181–5184
[15] Sivaramakrishnan R 2014 *International Journal of Innovative Research in Computer and Communication Engineering* **2** 2335–2339 ISSN 320 -9798
[16] Fulcher E and Patil S 1994 *Multichip Modules, 1994. Proceedings of the 1994 International Conference on* (IEEE) pp 572–577
[17] Niskanen S 2013
[18] Niskanen S 2009 *Helsinki University of Technology*

Synthesis and characterization of $La_{0.95}Ce_{0.05}CoO_3$ at different calcination temperatures

J Oviedo[1], C Caballero[2] and L C Sánchez[3]

[1] Departamento de Física y Electrónica, Universidad de Córdoba, Carrera 6 # 76 – 103 Montería, Córdoba, Colombia
[2] Departamento de Química, Universidad de Córdoba, Carrera 6 # 76 – 103 Montería, Córdoba, Colombia
[3] Grupo Avanzado de Materiales y Sistemas Complejos (GAMASCO), Universidad de Córdoba, Colombia.

E-mail: luiscarlos@correo.unicordoba.edu.co

Abstract. Cerium doped (5%) lanthanum cobaltites were synthesized in acid medium by using the sol-gel method. The obtained compounds were studied by using X-ray diffraction (XRD) and scanning electron microscopy (SEM). After the synthesis process, the samples were calcined at temperatures of 800 and 900 °C, observing changes in structural and morphological properties of annealed samples. Additionally, tests of catalytic activity were performed in the degradation reaction of methyl-thionine chloride, wherein the synthesized material was used as catalyst.

1. Introduction

The perovskites with metal transition, and possessing an ABO_3 formula, are generally well known as good oxidation catalysts [1,2], because they are cheaper than the noble metal supported catalysts. Perovskite type oxides can be adapted to create a wide family of catalysts, or varying the site A or B metal ion, or both. In fact, the catalytic activity of a perovskite may be modified by insertion of suitable transition metal ions. Partial substitution at site A with another cation by doping, can strongly affect the catalytic activity due to the stabilization of unusual oxidation states of the cation B and the simultaneous formation of structural defects created by such replacement. Structural defects are responsible for not only the catalytic activity but also the mobility of oxygen within the crystal lattice of the solid due to non-stoichiometry created by substitution. Lanthanum cobaltites both pure and doped have been recently used as catalyst. W. Y. Jung and S-S. Hong, investigated the photocatalytic activity for the decomposition of methyl orange on the $LaCoO_3$ perovskite-type oxides prepared at different conditions using microwave process [3]. H. Li et al, investigated effect of nickel on the catalytic performances of $LaCoO_3$ for CO oxidation, by partial sub-stitution of Co with Ni atoms in the framework or by deposition of NiO on the surface of $LaCoO_3$ [4]. Replacement of lanthanum in cobaltites increases the catalytic activity of these materials[5]. R. Newton et al, sinthetized perovskite-type oxides $La_{1-x}Ce_xCoO_3$ (x = 0, 0.05 and 0.10) evaluating the selective CO oxidation reaction [6].

Several synthetic methods have been developed wet route for preparing perovskite type oxides. Some of the most notable are sol-gel methods [7], Pechini-method also called polymerized citrate complex method [8], co-precipitation [9], microemulsion [10] and method hydro / solvothermal [11]. In the present study, Ce-doped lanthanum cobaltites ($La_{0.95}Ce_{0.05}CoO_3$) were synthesized, by sol-gel method with calcination temperatures of 800 and 900 °C, structural and morphological studies and catalytic activity tests of the obtained compounds were conducted.

2. Experimental

Materials synthesis was performed from nitrates considering the citrate/nitrate ratio, and using a reaction system that allowed add nitrates solutions on citric acid solution. For the synthesis of solids, 0.423 M solutions of lanthanum nitrate hexahydrate ($La(NO_3)_3 \bullet 6H_2O$), cobalt nitrate nonahydrate ($Co(NO_3)_2 \bullet 9H_2O$) and cerium nitrate hexahydrate ($Ce(NO_3)_3 \bullet 6H_2O$) in deionized water were prepared. Suitable volumes of the solutions were allowed to drip slowly with constant stirring with a solution of citric acid 0.133 M. After the dripping the temperature is maintained at 70 °C for two hours with stirring and reflux. The resulting solution was brought to an evaporation vessel to remove the solvent and allow the spontaneous combustion of the final complex. The obtained solids were crushed, sieved and calcined at 800 and 900 °C with heating rate 5 °C/min. X-ray diffraction (XRD) patterns were obtained with a Panalytical X.Pert PRO MPD equipment with CuKα1 radiation (λ=1.5418 Å) and the Scanning Electron Microscopy (SEM) was carried out with a JEOL/EO JSM-6490-LV equipment. The catalytic activity of the samples was conducted on the degradation reaction of methyl-thionine chloride, with 2 mg of solid, using a UV-Vis Pharo-300 spectrometer to quantify methyl-thionine chloride which was not degraded.

3. Results and discussion

In Figure 1, X-Ray diffraction patterns of calcined samples at 800 and 900 °C, adjusted by Rietveld refinement as implemented in the FULLPROF program are shown. The phase identification was carried out using XPowder and HighScore X'pert 2.1 programs, taking into account the data base PDF2-2004. The X-ray diffraction patterns were refined with the La-cations at 6*a (0, 0, 1/4)*, Co-cations at 6*b (0, 0, 0)* and oxygen anions at 18*e (0.4482, 0, 1/4)*. The parameters obtained from refinement and catalytic activity data are listed in Table 1. The identified phase for the samples corresponds to hexagonal perovskite structure and $R\bar{3}c$ (# 167) space group, without the presence of impurity phases. Similar results are reported by R. Newton et al [6] for $La_{1-x}Ce_xCoO_3$ (x = 0, 0.05 and 0.10) synthesized by a modified citrate method with calcination temperature of 700 °C. Furthermore authors obtained crystallites about 20 nm. Y Wen et al [12] found for $La_{1-x}Ce_xCoO_3$ (x = 0, 0.05, 0.1, 0.2, 0.3, 0.4), synthesized by citrate method with calcination temperature of 700 °C, rhombohedral structures for (x = 0, 0.05) and cubic structures when $x \geq 0.1$.

Lattice parameters and volume of the unit cell are not affected appreciably with the calcination temperature while the crystallite size increases with calcination temperature associated with increased crystallinity of the samples as shown in Table 1.

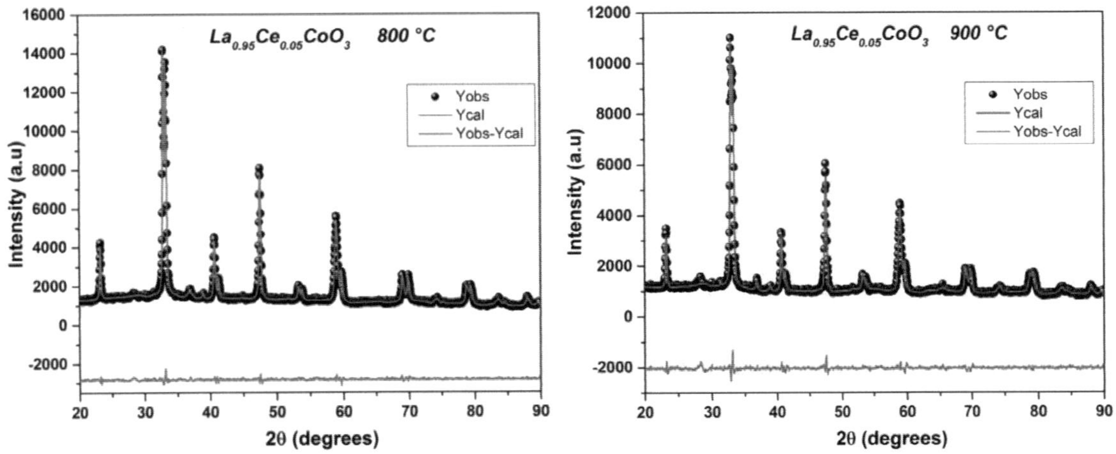

Figure 1. XRD patterns of $La_{0.95}Ce_{0.05}CoO_3$ samples calcined at 800 and 900 °C fitted by using Rietveld refinement.

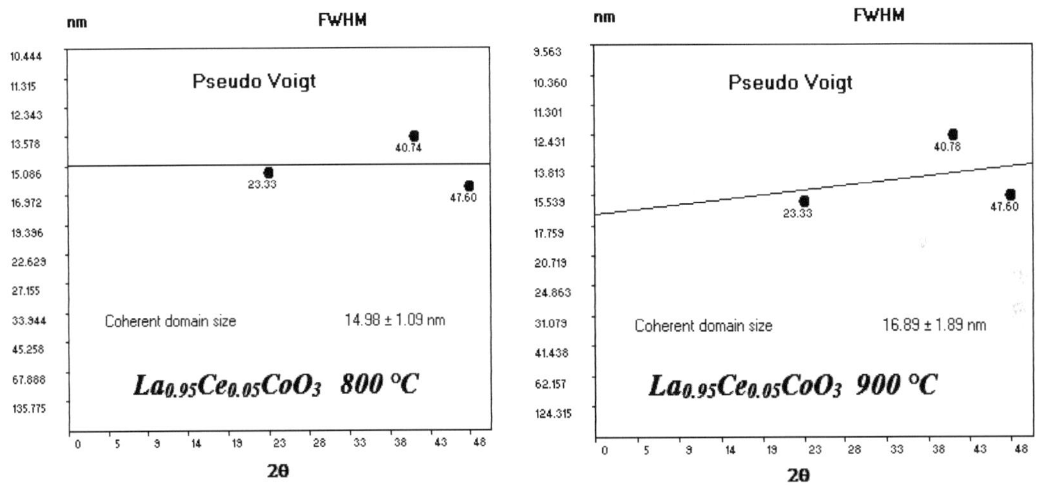

Figure 2. Williamson-Hall graphs considering Pseudo Voigt line profile of XRD patterns for samples calcined at 800 and 900 °C.

Table 1. Structural parameters obtained by Rietveld refinement and catalytic activity data.

Temperature (° C)	Structural parameters					Catalytic Activity
	a(Å)	*b*(Å)	*c*(Å)	*V*($Å^3$)	Crystallite size (nm)	% conversion
800	5.4415	5.4415	13.1117	336.2171	14.98±1.09	71.3
900	5.4388	5.4388	13.1241	242.9877	16.89±1.89	70.9

In Figure 3, SEM images of the calcined samples are shown. For these can be observed, in general different grain sizes (size distribution) with irregular crystallites and different dimensions possibly associated with the synthesis method. The obtained powders display aggregates of about 5–20 μm, which consist from crystallites with size of 14.98 nm and 16.89 nm for calcined samples at 800 and 900 °C respectively, as estimated by Williamson-Hall plots (Figure 2), in according to Ivanova S. et al [13]

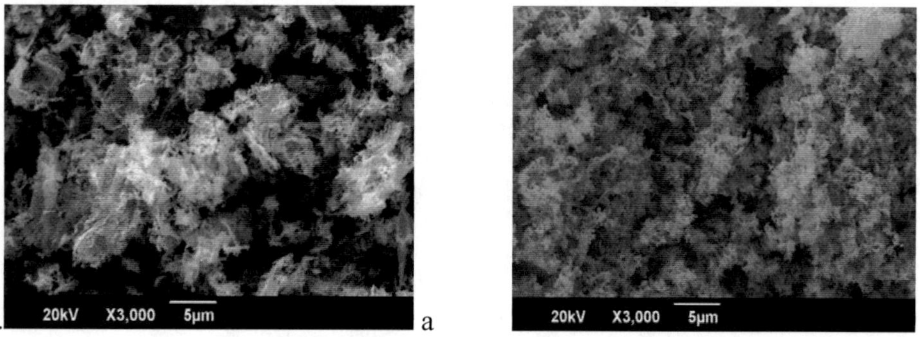

Figure 3. SEM images of $La_{0.95}Ce_{0.05}CoO_3$ calcined at (a) 800 °C and (b) 900 °C.

As result of the catalytic activity of samples in the degradation reaction of methyl-thionine chloride after 6 hours, were obtained conversion percentajes of 71.3 and 70.9 % for calcined samples at 800 and 900 °C, respectively (as shown in Table 1). It makes evident that the reaction can be catalyzed by Ce-doped lanthanum cobaltites, although is not observed a significant change in the degradation reaction (% conversion) with calcination temperature.

4. Conclusions

The calcination temperature of 800 and 900 °C of lanthanum cobaltites with Cerium at 5% substitution do not influences appreciably on the lattice parameters, but it does affect the crystallinity of the samples. Regarding the morphology of the samples it can be said that present agglomeration and there is no uniformity of grain sizes, fact associated with the synthesis method used. The catalytic activity of the samples in the degradation reaction of methyl-thionine chloride is not affected appreciably by the annealing treatment of solids, however, the lanthanum cobaltites with Cerium at 5% samples promotes the photocatalytic reaction.

References

[1] Spinicci R, Faticanti M, Marini P, De Rossi S, and Porta P 2003, *J. Mol. Catal. A: Chem* **197** 147.
[2] Farfan JC, Rodriguez JA, Fajardo F, Lopez EV, Tellez DAL, Roa-Rojas J 2009, *Physica B: Condensed Matter* **404:18** 2720
[3] Jung W Y and Hong S-S 2013, J. Ind. Eng. Chem. **19** 157
[4] Li H, et al 2016, Catal. Today. In Press.
[5] Deng J et al 2010, *Environ. Sci. Technol.* **44** 2618
[6] Newton R et al 2010, *Appl. Catal. A:Gen* **388** 216
[7] Cizauskaite S et al 2011, *Mater. Chem. Phys.* **125** 469
[8] Androulakis J, Migiakis P and Giapintzakis J 2004, *Appl. Phys. Lett* **84** 1099
[9] Singh C and Rakes M 2009, *Ind. J. Eng. Mater. Sci.* **16** 288
[10] Weidenkaff A 2004, *Adv. Eng. Mater.* **6** 709
[11] Pál E et al 2009, *Colloid Surface A* **340** 1
[12] Wen Y et al 2007, *Catal. Today.* **126** 400
[13] Ivanova S, Zhecheva E and Stoyanova R 2007, *J. Phys Chem. Sol.* **68** 168

Study of the suitability of a commercial hydroxyapatite powder to obtain sintered compacts for medical applications

C Palacio[1,3], D Jaramillo[1], S Correa[2] and M Arroyave[1]

[1]Departamento de Ciencias Físicas, Universidad Eafit, Medellin, Colombia.
[2]Grupo de Investigación en Bioingeniería, Universidad Eafit, Medellín, Colombia.
[3]Author to whom any correspondence should be addressed.

cpalac12@eafit.edu.co

Abstract. Hydroxyapatite (HA) is a material widely used by the medical community due to its Ca/P ratio is comparable to the Ca/P ratio of bones and teeth, which promotes osteoinduction and osteoconduction processes when in contact with bone tissue, either as volume piece or coating. This work focuses on studying the quality of the commercial HA powder MKnano-#MKN-HXAP-S12 μm, after processing, to obtain sintered compact discs with suitable physical and chemical characteristics for implants applications. The HA powder was processed through calcination, grinding, pressing and sintering to evaluate the effect of such as procedures in the compacts dics quality. The raw powder was characterized by laser diffraction, SEM, XRF, XRD, TGA and DSC while the characteristics of the obtained compact discs were determined by dilatometry and XRD to identify the sintering temperature range, constituent phases, the amorphous content and the crystallinity degree, parameters that allow determining their suitability for implants applications. Although, it was not possible to obtain sintered compacts with the suitable chemical composition and without fractures, this work allowed to identify the parameters that determine the suitability of a HA powder to obtain sintered compacts for medical applications, as well as the characterization protocol that allows the evaluation of such parameters.

1. Introduction

The Hydroxyapatite (HA) is a bioactive material because its Ca/P ratio is close to the Ca/P ratio of bones and teeth, this has turned the HA in a suitable material for clinical applications either as dense material or as coating over implants. The manufacturing processes of compact ceramics determine its properties and then influence the effectiveness of the processes that make use of these compacts [1]. The most used techniques to form compacts from HA powder are pressing and sintering [2], some others are: hot and cold pressing, hot isostatic and cold isostatic pressing; thermal plasma spraying; conventional, vacuum, spark and loose sintering; induction, arc and electron-beam melting, vacuum melting and casting; and co-precipitation technique.

The publications that study the in vitro behavior of HA sintered compacts report very different results about the microstructural, chemical and mechanical properties of these materials. It has been demonstrated that porous compacts favour the osteoconduction, but are less resistant to applied mechanical charges compared to dense compacts. On the other hand, the osteoinduction improves when the compact is amorphous but the durability of the compact is better as its crystallinity grade increases. Nevertheless, in most cases, the authors describe neither the manufacturing process of the dense compacts nor its final characteristics [3], [4]. At the same time, it turns out to be unusual that the processing parameters effects in the dense compacts characteristics are evaluated. This work focuses on studying the quality of the commercial HA powder MKnano-#MKN-HXAP-S12 μm, after processing, to obtain sintered compact discs with suitable physical and chemical characteristics for implants applications.

2. Materials and methods

The raw material used for the manufacture of the 2 in diameter HA discs consisted in the commercial HA powder #MKN-HXAP-S12 μm of the international mark MKnano, which reports for this powder 99 % of purity and an average particle size (APS) of 12 μm. The values of the powder processing conditions were taken from the reported values in the consulted literature for the realization of the state of the art: temperature and stabilization time during calcination 1000 ºC for 3 hours, compression force during pressing 500 kN, and temperature and stabilization time during sintering: 1100 ºC for 4 hours [5–7].

To evaluate the raw material and the effect of the identified processing condition, in the compact discs quality, the following characterization tests were realized: laser diffraction of the as-received and calcined powder in a Mastersizer 2000 and Cilas 930E, respectively; dilatometry in a Netzsch DIL 402 PC; Thermo-Gravimetry Analysis (TGA) and Differential Scanning Calorimetry (DSC) in a TA Instruments' Universal V4.5A; Scanning Electron Microscopy (SEM) of the MKnano powder particles in a JEOL JSM 6490 LV; X-Ray Fluorescence (XRF) spectrometry of the as-received and calcined powder in a ARL OPTIM' X WDXRF Spectrometer and PANalytical' AXIOS, respectively; and X-Ray Diffraction (XRD) in a PANalytical' XPert Powder and PANalytical' XPert Empyrean Serie II, for the phase identification and quantification by Rietveld and internal standard methods [8]. Fluorite, CaF_2 (32 % wt.), was used as internal standard, complying with the condition of compatibility with the analysis sample proposed by L. S. Zevin in 1979 [9].

3. Results and discussion

3.1. Particle size distribution

The APS increased and the Particle Size Distribution (PSD) decreased in amplitude after the calcination and grinding processes (see Figure 1), this is advantageous in relation to the homogeneity and densification that could be reached during the sintering process from powders [3]. In addition, a major APS implies smaller specific surface area, which can improve the powder manageability during compression.

3.2. Particles morfology

The particles morphology gives an idea about the powder synthesis process. Angular morphologies indicate that the material was melted and grinded and spheroidal morphologies indicate that it was

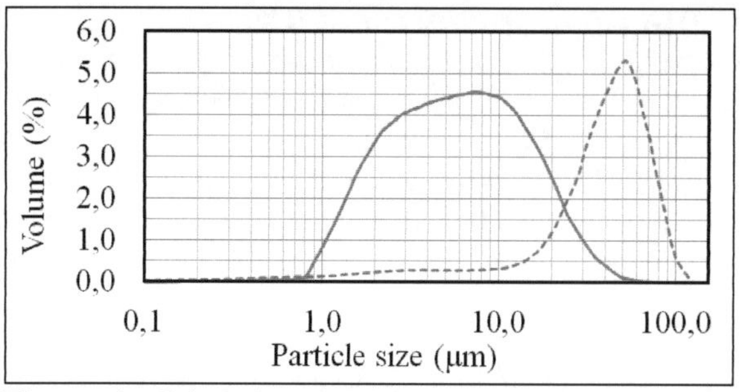

Figure 1. Grinded and calcined powder' PSD, APS 40 μm (dashed line) and as-received powder' PSD, APS 6. 2 μm (solid line).

Figure 2. SEM micrography (500X) of the MKnano powder particles sample.

agglomerated and sintered. The fact that the MKnano powder morphology (see Figure 2) is spheroidal turns out to be advantageous, since it is reported in the literature, that this morphology type is ideal for the controlled production of ceramic bodies [10].

3.3. Chemical composition

The powder composition analyzed by XRF (see table 1) indicates that the 99 % of purity reported by the provider is affected by a high content of potassium (K_2O) that stays after the calcination process, which goes to the detriment of the powder quality to be used in medical applications, considering the requests of obligatory fulfillment given by the standard ISO13779-6 [11] about impurities. The potassium content also prevents the determination of the Ca/P ratio.

3.4. Composition in phases and content of amorphous

From the XRD results presented in Figure 3, it was possible: i) to verify that MKnano powder consists of HA, as the provider informed, but also that the apatite is slightly crystalline and is accompanied by an amorphous material, quantified in 17.8 ± 5 % p/p by the internal standard method, and ii) to identify that both thermal treatments, calcination and sintering, drive to the transformation of certain quantity of this amorphous material in a phase constituted by potassium and phosphorus (KH_2PO_4).

Table 1. MKnano powder composition in oxides.

Oxides	As-received powder (% wt.)	Calcined powder (% wt.)
CaO	38.530	46.020
P_2O_5	24.380	34.750
K_2O	13.270	14.286
MgO	0.211	0.330
LOI	23.470	2.160

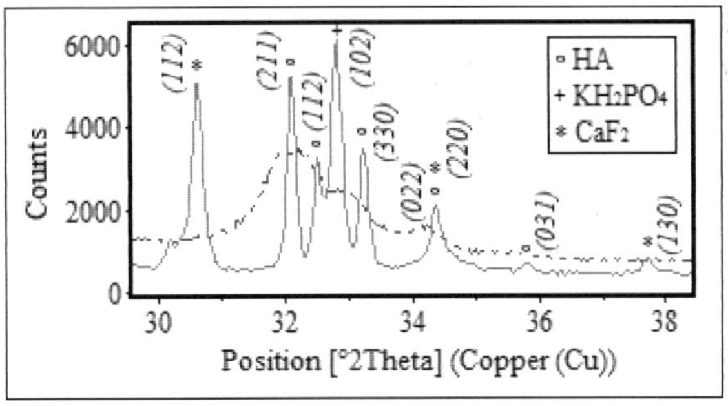

Figure 3. X-ray diffractogram of the grinded and calcined powder (solid line) and as-received powder (dashed line).

3.5. Thermal behaviour

The Figure 4 represents the curve of the powder thermal expansion parameter, which includes the material intrinsic thermal expansion, the contraction for sintering and phase transformations [12]. This information makes possible to identify that the sintering temperatures range for the MKnano powder is between 750 and 1170 °C because in this range takes place the greater contraction achieved by the evaluated compact, which is an indicator of the least volume that the sample achieves when applying

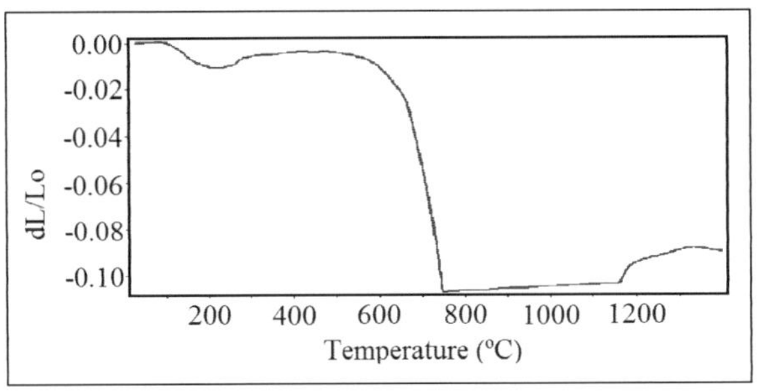

Figure 4. Curve of the HA compact sample thermal expansion.

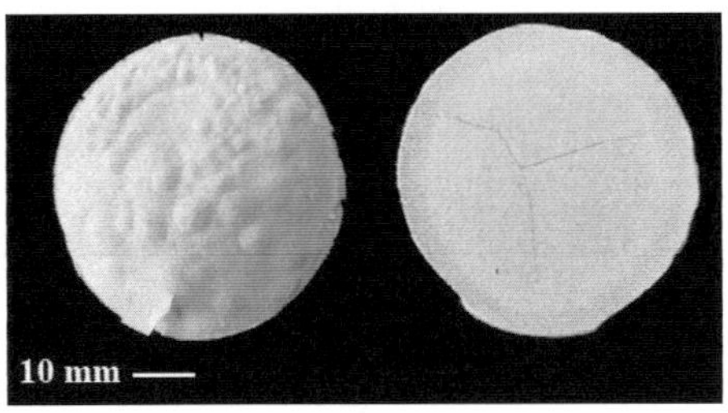

Figure 5. Sintered discs from calcined (right) and as-received powder (left).

thermal treatment. From the Figure 5 is possible to identify that: i) it is indispensable to calcine the powder before sintering, since phase transformations deform the HA compacts and ii) during cooling, there also happen physical phenomena that could fracture the discs.

4. Conclusions

ISO13779-6 standard [11] establishes two statements: i) the HA degradation phases in a powder for medical applications must not be bigger than 5 % wt., and ii) the maximum acceptable limit of traces is 30 mg/kg. This study evaluated the incorporation of a calcination treatment, prior to press and sintering, as a mechanism to remove the agglomeration agent based on potassium and in consequence to make the powder suitable for medical applications. Nevertheless, it was identified that both calcination and sintering thermal treatments conducted to partially transform the raw material in a crystalline phase constituted by potassium and phosphorus degrading in part the crystalline HA and affecting the Ca/P ratio ($1.66 \leq x \leq 1.71$) specified by the standard for powders usable in medical applications. In spite of the undesired chemical transformation of the MKnano powder and the fractures produced as consequence of the undesired chemical composition, this work allowed to identify the parameters that determine the suitability of a HA powder to obtain sintered compacts for medical applications, as well as the characterization protocol that allows the evaluation of such parameters. This procedure can be replicated in future works with commercial or homemade HA powders with the suitable and verified chemical characteristics to manufacture suitable sintered compacts of HA.

References

[1] Ozeki K, Aoki H and Fukui Y 2005 Effect of pH on crystallization of sputtered hydroxyapatite film under hydrothermal conditions at low temperature *J. Mater. Sci.* **40** 2837–42
[2] Bernache Assollant D and Bonnet JP 2005 Frittage: aspects physico - chimiques Partie 1: frittage en phase solide *Tech.l'ingénieur Fondam. en Chim.* **33** 1–20
[3] Rouahi M, Champion E, Gallet O, Jada A and Anselme K 2006 Physico - chemical characteristics and protein adsorption potential of hydroxyapatite particles *Colloids Surfaces Biointerfaces.* B **47** 10–9
[4] Nadaud N, Nanot M and Boch P 1994 Sintering and electrical properties of titania- and zirconia - containing In_2O_3-SnO_2 (ITO) ceramics *Commun Am Ceram Soc.* **77** 843–6
[5] Ding SJ, Ju CP and Chern Lin JH 1999 Immersion behavior of rf magnetron - assisted sputtered hydroxyapatite / titanium coatings in simulated body fluid *J. Biomed. Mater. Res.* **47** 551–63
[6] Hababi E, Harabi A, Foughali L, Chehlatt S, Zouai S and Mezahi FZ 2015 Grain growth in sintered natural hydroxyapatite *Acta Phys. Pol.* A **127** 1161–3
[7] Hong Z *et al.* 2007 Crystalline hydroxyapatite thin films produced at room temperature - an opposing radio frequency magnetron sputtering approach *Thin Solid Films* **515** 6773–80
[8] Westphal T, Füllmann T and Pöllmann H 2009 Rietveld quantification of amorphous with an internal standard - mathematical consequences of the experimental approach *JCPDS-ICDD Powder Diffr.* **24** 239–43
[9] Zevin LS 1979 Methods of mixing in quantitative x - ray phase analysis *J Appl Cryst.* **12** 582–4
[10] Jarby SE 2014 *Material product data sheet pure titanium thermal spray powders* (Pfäffikon: Oerlikon Metco) p 1
[11] International Organization for Standardization 2015 *ISO13779-6 Implants for surgery - hydroxyapatite: powders* (Geneva: ISO) ed 1
[12] Abdel Fattah WI and Nour FA 1993 Thermal expansion application to assess calcination of bovine hydroxyapatite *Thermochim. Acta* C **218** 465–75

SYNTHESIS OF ALUMINA THIN FILMS USING REACTIVE MAGNETRON SPUTTERING METHOD

G Angarita, C Palacio, M Trujillo, M Arroyave

Universidad EAFIT, Applied Electromagnetism Research Group, Cr49, No.7 sur 50, Medellin, Colombia

Email: langari3@eafit.edu.co

Abstract. Alumina (Al_2O_3) thin films were deposited on Si (100) by Magnetron Sputtering in reactive conditions between an aluminium target and oxygen 99.99% pure. The plasma was formed employing Argon with an R.F power of 100 W, the dwelling time was 3 hours. 4 samples were produced with temperatures between 350 and 400 °C in the substrate by using an oxygen flow of 2 and 8 sccm, the remaining parameters of the process were fixed. The coatings and substrates were characterized using Atomic Force Microscopy (AFM), Scanning Electron Microscopy (SEM), X-ray diffraction (XRD) and Energy Dispersive Spectroscopy (EDS) in order to compare their properties before and after deposition. The films thicknesses were between 47 and 70 nm. The results show that at high oxygen flow the alumina structure prevails in the coatings while at lower oxygen flow only aluminum is deposited in the coatings. It was shown that the temperature increases grain size and roughness while decreasing the thicknesses of the coatings.

1. Introduction

Aluminum oxide or Alumina (Al_2O_3) is an outstanding ceramic material which, due to its several excellent physical and chemical properties like high electric resistivity, low thermal conductivity, high melting temperature, high hardness, refractive index of 1.7 [1], high abrasion and oxidation resistance, finds application in cutting tools, optoelectronics devices, diffusion and thermal barriers [2]. Al_2O_3 has several metastable phases structures where the most reported are: cubic γ and η phases, the monoclinic θ phase, the hexagonal χ phase, the orthorhombic κ phase and the δ phase that can be tetragonal or orthorhombic [3]. Al_2O_3 only has one stable phase (α-Al_2O_3) with a trigonal hR30 structure, melting point of 2043 °C, and a relatively high hardness (21 GPa) [4]. The metastable phases appear when alumina is being heated in thermal processes like annealing; the ranges in which the metastable phases are formed depend upon the composition and structure of the initial material, where the stable α-Al_2O_3 is formed at about 650 °C [4]. A large number of techniques have been reported for the deposition of amorphous and crystalline alumina in the literature, such as magnetron sputtering, ion assisted deposition, pulsed laser deposition, electron beam evaporation, chemical vapor deposition and sol-gel deposition [5]. This work focuses in the deposition of crystalline alumina from reactive magnetron sputtering technique by controlling the substrate temperature and oxygen flow.

2. Materials and Methods

Al_2O_3 films were deposited by reactive magnetron sputtering on p-type silicon (100) substrates (15 x 15 mm^2) with a native oxide layer on top of around 10 nm thick. The substrates were ultrasonically cleaned in isopropanol for 10 minutes and dried with nitrogen gas. A 99.9995 % pure Aluminum target with 50.80 mm in diameter and 6.35 mm thick was used. Prior to the deposition, the chamber was evacuated with a turbomolecular pump to a background pressure of 4.2 x 10^{-6} Torr. Ultra high purity (99.9999%) argon and oxygen gasses (Oxygen/Argon ratio of 1/5) were then introduced into the chamber using a mass flow regulator system. The desired working pressure of 6 x 10^{-3} Torr was obtained by regulating the pumping speed, which is controlled by a throttle valve in the path of high vacuum line connecting the chamber to the turbomolecular pump. Before deposition, a pre-sputtering was performed at 100 W RF power for 10 minutes in order to clean the surface of the Al target. The deposition time for all coatings was fixed at 3 hours and the influence of the substrate temperature and oxygen flow was evaluated by employing a factorial design using these two parameters at two different levels chosen according to the ranges found in the literature (2^2 factorial design) [6]. Oxygen flow varied from 2 to 8 sccm and the temperature levels were chosen at 350 and 400ºC.

The morphological properties and deposition rate of the films were evaluated using Atomic Force Microscopy (AFM) with a NANOSURF EasyScan 2 device using a CT170 tip in noncontact mode. The deposition rate of the coatings was measured by marking a line on the substrates with an alcohol soluble marker in order to create step heights that were then measured by AFM. The roughness measurements were made over an area of 50 x 50 μm^2 and the grain size measurements were made over 5 x 5 μm^2. Meanwhile, Scanning Electron Microscopy (SEM) was used to evaluate the morphology of the coatings, and Energy Disperse Spectroscopy X-Ray (EDS) was used to qualitatively observe the different elements in the deposited coatings, these two techniques were accomplished by using a JEOL JSM-6701F Field Emission Scanning Electron Microscope (FESEM). X-Ray Diffraction (XRD) was used to study the crystalline phases of the coatings with a PANalytical Diffractometer Empyrean device using a copper anode ($K\alpha$ = 1.5405 Å and $K\beta$ = 1.5444 Å).

3. Results and discussion

3.1. Scanning Electron microscopy

Figure 1. shows a micrograph taken by FE-SEM at 270,000 magnifications of the deposited films, where a granular structure can be seen, this structure is typical of coatings deposited by sputtering.

Figure 1. FE-SEM micrograph of a coating deposited at 350ºC and 8sccm.

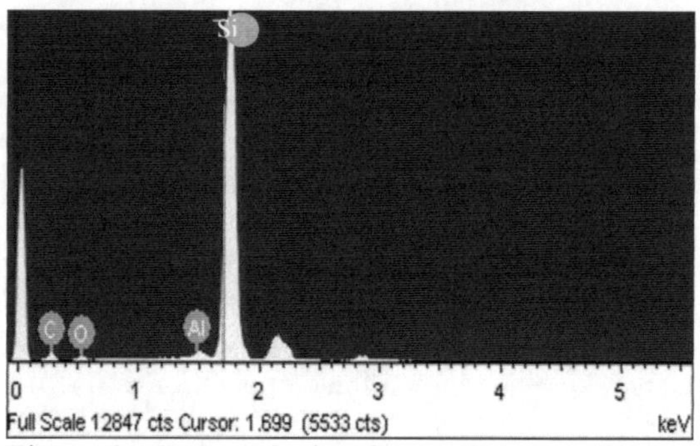

Figure 2. EDS analysis of a coating deposited at 350ºC and 8sccm.

EDS graph (Figure 2) shows the presence of aluminum, oxygen, silicon and carbon in the samples, which proves that aluminum in presence of oxygen was deposited; the silicon is attributed to the substrate and carbon to the impurities of the environment.

3.2. Atomic force microscopy

The measured thickness and calculated deposition rate of the films can be observed in Figure 3, which reveals that the higher the temperature, the more material is deposited, this can be attributed to a greater affinity of the aluminum and oxygen adatoms [7] to the substrate surface at high temperatures.

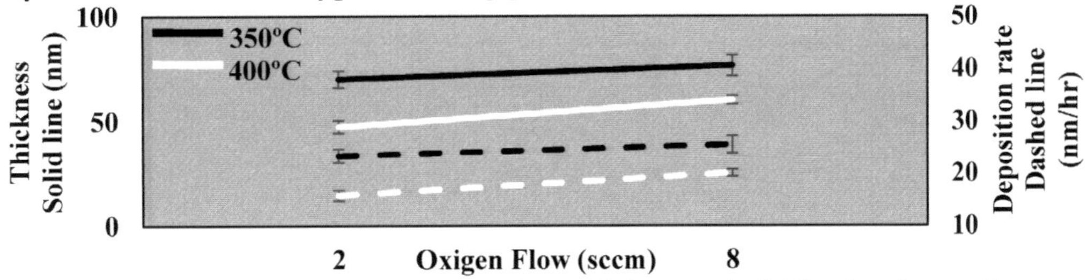

Figure 3. Oxygen and temperature influence in the thicknesses (solid lines) and sputtering rates (dashed lines) of the coatings.

Figure 4 shows the morphology of the coating, were a granular structure with some cluster type grain agglomerations can be seen, this structure is typical of coatings deposited by sputtering where the films growth is of columnar nature.

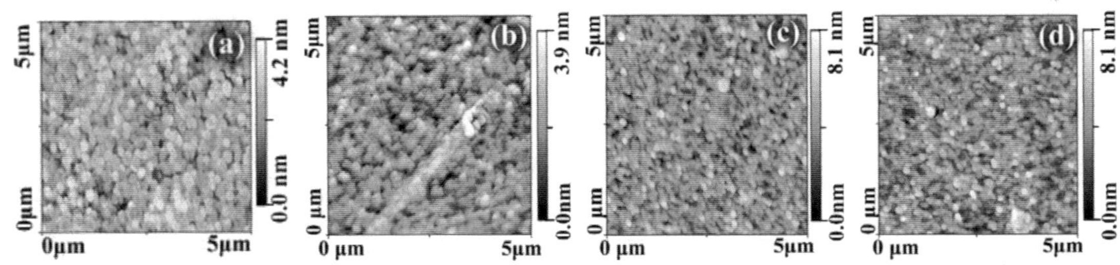

Figure 4. AFM micrographs of deposited coatings at: (a) 350°C and 2 sccm, (b) 350°C and 8 sccm, (c) 400°C and 2 sccm and (d) 400°C and 8 sccm.

Figure 5 shows the grain sizes and roughness of the coatings. It can be seen that the grain size and roughness decreases when the temperature increases, this behaviour can be explained by the increase of the nucleation sites due to a larger travel time of the adatoms in the substrate surface which in turn produces higher quantity of grains with small grain sizes in the film [7].

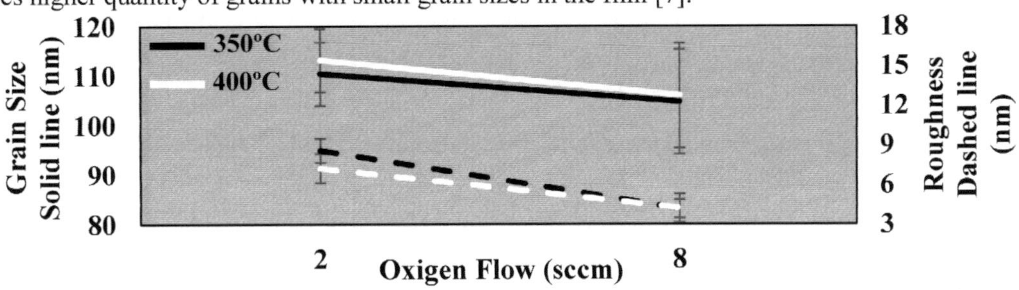

Figure 5. Oxygen flow and substrate temperature influence in the grain size (solid line) and roughness (dashed line) of the deposited films.

3.3. X-Ray diffraction

The normalized XRD spectra of the films is shown in Figure 6. The patterns used for the identification of the XRD peaks belong to the ICCD and COD databases and have the following reference codes: 00-026-0031 (Al_2O_3) [8], 96-901-2003 (Al) [9] and 96-901-2602 (SiO_2) [10].

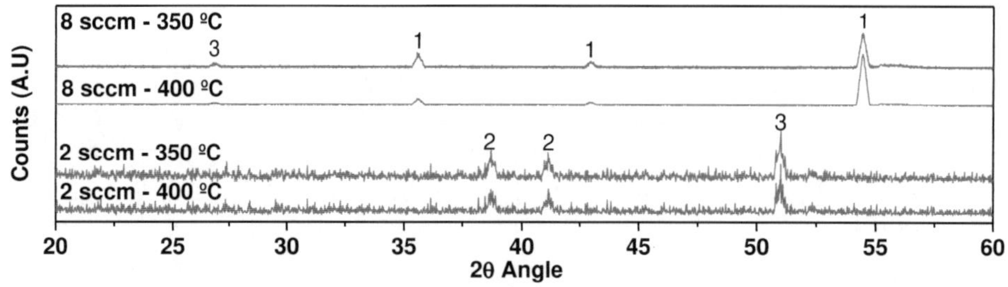

Figure 6. XRD spectra of the synthetized coatings, 1) κ-Al_2O_3 [8], 2) Al [9], 3) SiO_2 [10].

Figure 6 shows that under high oxygen flow (8 sccm) orthorhombic alumina (κ-Al_2O_3) was formed in the films. At low oxygen flow (2 sccm) only aluminum and a greater noise in the XRD can be observed, this is attributed to the low content of oxygen in the chamber, which is not enough to form a stoichiometric alumina structure in the entire film. The high noise can be attributed to a film rich in aluminum with a somewhat amorphous structure from the little reaction between aluminum and oxygen. In the other hand, the temperature at which the experiments were carried out was not enough to make an important change in the XRD spectra of the coatings.

4. Conclusions

Orthorhombic alumina (κ-Al_2O_3) films were produced by reactive RF magnetron sputtering with grain sizes ranging from 100 and 115 nm. The oxygen flow plays a crucial role in synthesizing alumina; it was found that at 2 sccm only Aluminium was deposited whereas at 8 sccm the quantity of oxygen was enough to form Al_2O_3. In addition, it was found that at higher oxygen flow rates, the coatings had an increase in their grain size and roughness.

It was found that the substrate temperature was not sufficient to cause changes in the crystalline structure of the material; however, there were found significant decreases in the thickness of the coatings at higher temperatures as well as increases in their roughness and grain size.

References

[1] Wypych G 2010 Physical and Chemical Properties *Handbook of Fillers (3rd Edition).*

[2] Khanna A and Bhat D G 2006 Nanocrystalline gamma alumina coatings by inverted cylindrical magnetron sputtering *Surf. Coatings Technol.* **201** 168–73.

[3] Levin I and Brandon D 1998 Metastable Alumina Polymorphs: Crystal Structures and Transition Sequences *J. Am. Ceram. Soc.* **81** 1995–2012.

[4] Zhang S and Ali N 2007 The Physical, Mechanical, and Thermal Properties of Hard Materials *Nanocomposite thin films and coatings : processing, properties and performance* (Imperial College Press) p 615.

[5] Zhang X, Zhu J, Zhang L, Kishimoto K, Du S and Yin X 2013 Crystallization of alumina films deposited by reactive magnetron sputtering with resputtering technique at low temperature *Surf. Coatings Technol.* **228** S393–6.

[6] Box G E P, Hunter J S and Hunter W G 2005 *Statistics for Experilllenters: Design, Innovation, and Discovery* (New Jersey: Wiley-Interscience).

[7] Mattox D M 1998 *HANDBOOK OF PHYSICAL VAPOR DEPOSITION (PVD): Film Formation, Adhesion , Surface preparation and contamination control* (Noyes Publications).

[8] Okumiya M and Yamaguchi G 1971 The Crystal Structure of κ′-Al 2 O 3 , the New Intermediate Phase *Bull. Chem. Soc. Jpn.* **44** 1567–70.

[9] Fortes A D, Wood I G, Vočadlo L, Knight K S, Marshall W G, Tucker M G and Fernandez -Alonso F 2009 Phase behaviour and thermoelastic properties of perdeuterated ammonia hydrate and ice polymorphs from 0 to 2GPa *J. Appl. Crystallogr.* **42** 846–66.

[10] Hazen R M, Finger L W, Hemley R J and Mao H K 1989 High-pressure crystal chemistry and amorphization of quartz *Solid State Commun.* **72** 507–11.

Synthesis of Silver nanoparticles (AgNPs) with Antibacterial Activity.

Gloria E. Campillo[1], Ederley Vélez[1], Gladis Morales[2], César Hincapié[1], Jaime Osorio[3], Oscar Arnache[3], José Ignacio Uribe[3], Franklin Jaramillo[4]

[1] Departamento de Ciencias Básicas, Grupo de Materiales Nanoestructurados y Biomodelación – MATBIOM, Universidad de Medellín, Colombia

[2] Facultad de Ingenierías. Grupo de investigaciones y Mediciones Ambientales – GEMA, Universidad de Medellín. Medellín, Colombia

[3] Instituto de Física. Facultad de Ciencias Exactas y Naturales. Universidad de Antioquia, Calle 70 No. 52-21, A.A. 1226, Medellín, Colombia

[4] Centro de investigación, innovación y desarrollo de materiales – CIDEMAT, Universidad de Antioquia UdeA, Calle 70 No 52-21, Medellín, Colombia.

E-mail: gecampillo@udem.edu.co

Abstract. The synthesis of nanomaterials is currently one of the most active in nanoscience branches; especially those help improve the human quality life. Silver nanoparticles (AgNPs) are an example of this as it is known to have inhibitory and bactericidal effects. In this work, we report the synthesis of silver nanoparticles by chemical reduction method of silver nitrate ($AgNO_3$) from aqueous solution, using a mix of polivinyl pyrrolidone (PVP) – *Aloe Vera* as reducing agent and for stabilization and control of particle size. Silver nanoparticles obtained were characterized by Scanning Electron Microscopy (SEM), UV–visible spectroscopy and measurements using Zetasizer Nano ZS were applied to size estimation. The existence of surface plasmon resonance peak at λ_{max} ~ 420 nm is evidence of silver nanoparticles formation. It was possible to standardize an appropriate protocol for the evaluation of bactericidal activity of the nanoparticles, for mesophilic microorganisms. Bactericidal activity above 90% against these kinds of bacteria was demonstrated.

Key words: Silver Nanoparticules, Antibacterial Activity, nanomaterial synthesis, mesophilic bacteria.

INTRODUCTION

Since ancient times it is known that silver has antibacterial properties, it has been used in biomedical applications, water and air purification, food production, cosmetics, clothing, and numerous household products [1]. Silver nanoparticles might exhibit additional antimicrobial capabilities not exerted by ionic silver, because its small size and large surface to volume ratios, which lead to both chemical and physical differences in their properties compared with their bulk counterparts [2]. AgNPs can be produced with various sizes and shapes depending on the fabrication method, among which the most widely used is the method of chemical reduction [3]. The aim of this study was to evaluate the antibacterial activity of AgNPs which were synthesized by chemical reduction method of silver nitrate ($AgNO_3$) in water, using polivinylpyrrolidone (PVP) as reducing agent and *Aloe Vera* for stabilization and size control. The characterization of AgNPs was carried out by different techniques and antibacterial activity against mesophilic microorganisms was measured by number of bacteria colonies in Agar Plate Count.

EXPERIMENTAL

- **Preparation of Silver nanoparticles (AgNPs).**

The AgNPs were prepared by chemical reduction of an aqueous solution 12mM of $AgNO_3$. At 70 mL of this solution is added PVP (keeping the molar ratio of the repeating unit of PVP and Ag equal to 34) and 21 mL of *Aloe Vera*, the whole reaction was carried out in argon atmosphere. The mixture was stirred in ultrasound at room temperature for 45 min, and then heated 2°C/min to reach 80°C and left for 2 hours to obtaining a translucent solution with small suspended particles must be removed by simple filtration.
For purposes of comparison was done a synthesis with Aloe Vera only, to identify whether this natural extract may be used with dual function as a reducing agent and stabilizer. However, although using only Aloe Vera, a monodisperse suspension with an average particle size of 82.61 nm was obtained, this solution had no stability over time. Therefore, it was possible establish the size modulator and stabilization capacity of PVP, indicating that is necessary to use a PVP-*Aloe Vera* mixture to favor the formation of small nanoparticles and stabilized over time.

- **Characterization.**

The particle size analysis was performed 24 hours after synthesis; the samples were studied by use dynamic light scattering measurements a Nanozeta S (Malvern), UV–vis absorption spectroscopy (UV-Visible Cary-100 VARIAN) in the wavelength range from 200 to 900 nm. The size and morphology of silver nanoparticles were observed by Scanning Electron Microscopy (SEM) using a JEOL JSM-6490LV. Samples for SEM measurements were suspended in ethanol and ultrasonically dispersed and by Trasmission Electron Microscopy (TEM) using a Tecnai F20 Super Twin TMP with Field emission source, resolution of 0.1 nm at 200 kV, 1.0 maximum magnification TEM MX camera GATAN US 1000XP-P.

- **Antibacterial activity.**

First we proceeded to prepare the inoculum for which was available fresh Rumen, the nutrient solution was prepared according to Siegert and Banks [4], 1 g of Rumen per liter of solution. It was cultured with 10 mL of the inoculated solution and stirred for 24 hours. Nutrient broth solution according to the manufacturer's recommendation (8g/L) was prepared and spread onto Agar Plate Count and incubated at 36- 37°C for 6 h. After incubation, the content of microorganisms was evaluated according to the scale McFarland (($1.7x10^8$ CFU/mL). To examine the bactericidal activity of AgNPs, different concentrations of AgNPs solution (0.1, 1, 10, 20, 40, 60, 80 and 100%) was used. An Agar Plate Count solution was used as nutrient medium, which was sterilized at 20 pounds of pressure for 30 minutes in autoclave GEMMYCO SA 232 mark.
In a series of Petri dishes previously sterilized it was added a volume of silver solution such that 10% remained in the volume concentration of agar to each system. When solidify the agar, it was cultured with 0.1mL of inoculum and incubated at 37°C (corresponding mesophilic microorganisms). After incubation, the number of colonies grown on the agar was counted [5]. The agar without nanoparticles and inoculated was considered as positive control, while a negative control, consisted of agar with nanoparticles but uninoculated.

RESULTS AND DISCUSSION

- **Synthesis and Characterization**

Due to the surface Plasmon resonance effect, is possible indicating of metal nanoparticles presence because the small metal nanoparticles exhibit the absorption of visible electromagnetic waves by the collective oscillation of conduction electrons at the surface [6]. UV-Vis spectroscopy was used to identify the presence of AgNPs, Figure

1a) shows the UV-Vis absorbance spectrum of AgNPs obtained, showing a peak associated with surface plasmon resonance at λ_{max} ~ 420 nm, although it can be turned by the fabrication technique [7]. Others works have reported a maximum peak between 405-418 nm [8,9]. The size distribution by intensity of synthesized AgNPs, is showed in Figure 1b, and confirms the presence of nanoparticles with size less than 20nm, as well as the presence of nanoparticles of greater size but, in minimum quantities. Therefore, it is clear that in this synthesis of AgNPs is obtained a polydisperse suspension, possibly due to the effect of the mixture of both stabilizers and a possible antioxidant activity of PVP that prevents aggregation of the nanoparticles in the system.

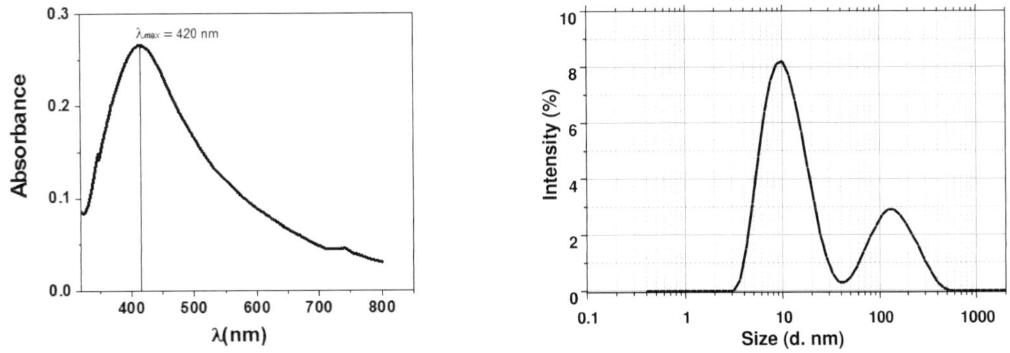

a) b)

Figure 1: a) UV-vis spectrum and b) particle size distribution analysis of AgNPs

The shape and size distribution of AgNPs were characterized by SEM and TEM micrographies, which are shown in Figure 2. It confirms the presence of nanoparticles with a size between 10 nm at 130 nm, we used *Aloe Vera* as surfactant to avoid nuclei aggregation in order to decrease the total surface energy. The use of substances that lead to steric repulsion between individuals is one way to prevent nanoparticles from aggregation [10].

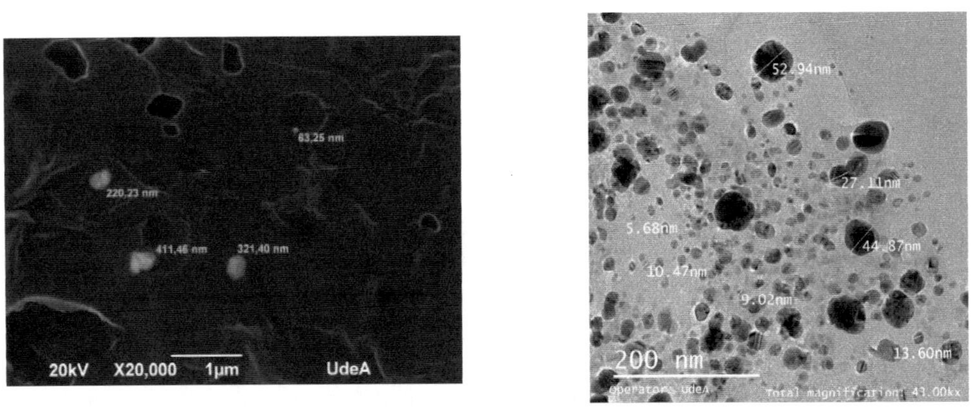

a) b)

Figure 2: a) Micrographs SEM and b) TEM of AgNPs with *Aloe Vera*-PVP

- **Antibacterial activity**

The antibacterial activity of AgNPs against a mesophilic microorganism, the *Kocuria varians* a gram-positive cocci, was evaluated by triplicate in agar plates. Bactericidal activity was shown at higher concentration (20, 40, 60, 80 and 100%) other dilutions (0.1, 1 and 10%) didn't have antibacterial activity. Figure 3 shows photographs of the antibacterial test results, the lowest concentration that completely inhibited bacterial growth was 20%.

Figure 3: Photographs of the antibacterial test results of AgNPs at 100, 80, 60, 40 and 20%, made by triplicate.

CONCLUSIONS

Chemical reduction method was used to synthesize the AgNPs whit an average size of 20 nm, form silver nitrate ($AgNO_3$) in water and using polivinylpyrrolidone (PVP) as reducing agent and *Aloe Vera* for stabilization and size control with an average size of ~20nm. These AgNPs allowed standardizing an appropriate protocol for the evaluation of their bactericidal activity on mesophilic microorganisms. It was found that for all dilutions evaluated the nanoparticles exhibit bactericidal activity after 24 hours and the lowest concentration that completely inhibited bacterial growth was 20%.

ACKNOWLEDGMENTS

This work was supported by Project 764 of Vicerrectoría de Investigaciones of Universidad de Medellín and the Solid State Group – GES, Sustainability Strategy 2016-2017 CODI – Universidad de Antioquia. The authors also gratefully acknowledge to Centro de Laboratorios of Universidad de Medellín.

REFERENCES

[1] Catalina Marambio-Jones, Eric M. V. Hoek. A review of the antibacterial effects of silver nanomaterials and potential implications for human health and the environment. *J. Nanopart. Res.* (**2010**) 12:1531–1551.

[2] Mohamed A. Shenashen, Sherif A. El-Safty, and Emad A. Elshehy. Synthesis, Morphological Control, and Properties of Silver Nanoparticles in Potential Applications. *Part. Part. Syst. Charact.* (**2014**), 31: 296- 316.

[3] C.I. Contescu, K. Putyera, Dekker encyclopedia of nanoscience and nanotechnology, vol. I-VI. ISBN 13 978-0-8493-9639-7. (**2009**).

[4] Irene Siegert, Charles Banks, The effect of volatile fatty acid additions on the anaerobic digestion of cellulose and glucose in batch reactors. *Process Biochemistry*, (**2005**), 40, :3412–3418.

[5] Wang, J.-X., L.-X. Wen, Z.-H. Wang, and J.-F. Chen. Immobilization of silver on hollow silica nanospheres and nanotubes and their antibacterial effects. *Mater. Chem. Phys.* (**2006**) 96: 90-976

[6] S. Link, M. A. El-Sayed, Shape and size dependence of radiative, nonradiative, and photothermal properties of gold nanocrystals. *Int. Rev. Phys. Chem.* (**2000**), 19: 409–453.

[7] M. Kuisma, A. Sakko, T. P. Rossi, A. H. Larsen, J. Enkovaara, L. Lehtovaara, and T. T. Rantala. Localized surface plasmon resonance in silver nanoparticles: Atomistic first-principles time-dependent density-functional theory calculations, *PHYSICAL REVIEW B* 91, 115431 (**2015**)

[8] Guzman M.G., Dille J., Godet S., Synthesis of silver nanoparticles by chemical reduction method and their antibacterial activity, *Int. J. Chem. Biol. Eng.*, (**2009**) 2, 104-111.

[9] E. Saion, E. Gharibshahi, K. Naghavi. Size-Controlled and Optical Properties of Monodispersed Silver Nanoparticles Synthesized by the Radiolytic Reduction Method. *Int. J. Mol. Sci.* (**2013**) 14, 7880 -7896.

[10] Le T. T. T., Fribourg-Blanc E., Dang M. C. Synthesis and optical properties of copper nanoparticles prepared by a chemical reduction method. *Adv. Nat. Sci.: Nanosci. Nanotechnol.* 2 (**2011**) 1–6.

IOP Conf. Series: Journal of Physics: Conf. Series **850** (2017) 012024 doi:10.1088/1742-6596/850/1/012024

Transfer matrix method in systems with topological insulators

J C Granada E [1] **and D F Rojas** [1]

[1] Departamento de Física, Universidad del Valle, AA 25360 Cali, Colombia

juan.granada@correounivalle.edu.co

Abstract. In the frame of axion electrodynamics a 4×4 transfer matrix method was developed in order to describe the electrodynamic properties of systems containing topological insulators, which are characterized by non trivial topological properties related to time reversal symmetry. Analytical expressions were obtained for the components of the transfer matrix for the interface between a conventional dielectric and a topological insulator, which generalize the relations obtained in conventional electrodynamics for the description of the transmission of TE and TM polarized modes. It was shown that such matrix couples the TE and TM modes, where the terms which are related to such coupling are linear with respect to the fine structure constant. Analytical expressions were obtained for the changes induced by the non trivial topology in the reflection coefficients from TM and TE incident waves and for the rotation of the polarization planes of reflected and transmitted waves.

1. Introduction

Topological insulators (TI) [1-6] belong to a class of solids that have strong spin-orbit coupling and are insulating in the bulk, but have an odd number of gapless conducting surface states protected against time-reversal invariant disorder. When the symmetry is broken in a 3D TI (by proximity to a magnetic material or by an externally applied magnetic field), a topological magnetoelectric effect arises, which is associated to a polarization charge induced by a magnetic induction $\boldsymbol{B}$ and a magnetization induced by an electric field $\boldsymbol{E}$. In the context of classical electrodynamics, the description of the topological magnetoelectric effect contains a term of the form $\theta(\boldsymbol{E}\cdot\boldsymbol{B})$ (where the topological axion parameter θ takes values that are integer multiples of π), leading to the following modified constitutive relations between the electric displacement $\boldsymbol{D}$, the magnetic field $\boldsymbol{H}$ and the fields $\boldsymbol{E}$, $\boldsymbol{B}$ [7]:

$$\boldsymbol{D} = \varepsilon\boldsymbol{E} - \varepsilon_0\alpha\frac{\theta}{\pi}c\boldsymbol{B}\,, \quad c\boldsymbol{H} = \frac{c\boldsymbol{B}}{\mu} + \alpha\frac{\theta}{\pi}\frac{\boldsymbol{E}}{\mu_0}\,. \tag{1}$$

In the above expressions ε, ε_0 and μ, μ_0 are the dielectric permittivities and magnetic permeabilities of the topological insulator and vacuum, respectively and α is the fine-structure constant.

The axion coupling results in the modification of the usual laws of electromagnetic wave propagation [8]: in particular, light reflected off an interface where the topological parameter θ jumps experiences a nontrivial Kerr rotation of its polarization and the light transmitted through the interface also undergoes a nontrivial Faraday rotation; in both cases the rotation of the polarization plane induced by the non trivial topology is a linear function of the jump of the topological term. The accumulation of the polarization rotation effect occurring in multilayer walls made of alternating layers of a topological insulator and a normal insulator can enhance the Casimir repulsion between two parallel multilayer walls [9, 10]. Additionally, axion coupling introduces changes in the spectra of collective excitations arising at topologically nontrivial interfaces, such as plasmons and surface plasmon-polaritons in nanostructured topological insulators (thin films, multilayers, and nanoribbons) [11-14].

Transfer matrix formalism [15, 16] has been used extensively for the description of wave propagation in a wide range of physical systems, including photonic crystals and metamaterials. For systems containing topological insulators some results have been reported for the calculation of the reflection coefficients of an electromagnetic wave reflecting on the interface with a jump of the topological term [9, 10]. It is the aim of this communication to give a detailed derivation of the analytical expressions of the transfer matrix for the interface between a conventional dielectric and a topological insulator, in order to generalize the relations obtained in conventional electrodynamics for the description of the transmission of TE and TM polarized modes.

2. Model and general relations

The system consists of two semi-infinite homogeneous isotropic TI media with dielectric permittivities $\varepsilon^{(1)}$, $\varepsilon^{(2)}$, magnetic permeabilities $\mu^{(1)}, \mu^{(2)}$, topological parameters $\theta^{(1)}$, $\theta^{(2)}$, and occupying the regions $z<0$, $z>0$, respectively. Because of the homogeneity of the system in the x and y directions, the y-components of the electric ($\boldsymbol{E}$) and magnetic ($\boldsymbol{H}$) fields at the different regions can be taken in the form (where the superscripts +, − indicate the forward and backward directions, respectively)

$$\begin{pmatrix} E_y^{(1)}(x,z,t) \\ H_y^{(1)}(x,z,t) \end{pmatrix} = \begin{pmatrix} E_y^{1+} \\ H_y^{1+} \end{pmatrix} \exp\left[i\varphi^{(1-)}(x,z,t)\right] + \begin{pmatrix} E_y^{1-} \\ H_y^{1-} \end{pmatrix} \exp\left[i\varphi^{(1+)}(x,z,t)\right],\ z<0 \tag{2a}$$

$$\begin{pmatrix} E_y^{(2)}(x,z,t) \\ H_y^{(2)}(x,z,t) \end{pmatrix} = \begin{pmatrix} E_y^{2+} \\ H_y^{2+} \end{pmatrix} \exp\left[i\varphi^{(2+)}(x,z,t)\right] + \begin{pmatrix} E_y^{2-} \\ H_y^{2-} \end{pmatrix} \exp\left[i\varphi^{(2-)}(x,z,t)\right],\ z>0. \tag{2b}$$

Here $\varphi^{(i\pm)}(x,z,t) = k_x x \pm k_z^{(i)} z - \omega t$, $k_z^{(j)} = \sqrt{\frac{\omega^2}{c^2}\varepsilon^{(j)}\mu^{(j)} - k_x^2}$, i=1, 2. The remaining components of the fields can be found with the aid of conventional Lorentz-Maxwell equations.

Note that by matching the phases of the plane waves (2a, 2b) at z=0, it follows that the laws of reflection and refraction remain valid even at the interface where the topological parameter θ jumps [8]. The continuity of the components E_x, E_y, H_x, D_z at the surface z=0 leads to the following relation between the amplitudes of the electric and magnetic fields:

$$\begin{pmatrix} 1 & 1 & 0 & 0 \\ -\frac{k_z^{(1)}}{\mu^{(1)}} & \frac{k_z^{(1)}}{\mu^{(1)}} & -\frac{\alpha\theta^{(1)}}{\pi}\sqrt{\frac{\mu_0\varepsilon^{(1)}}{\varepsilon_0\mu^{(1)}}}\frac{k_z^{(1)}}{\mu^{(1)}} & \frac{\alpha\theta^{(1)}}{\pi}\sqrt{\frac{\mu_0\varepsilon^{(1)}}{\varepsilon_0\mu^{(1)}}}\frac{k_z^{(1)}}{\mu^{(1)}} \\ \frac{\alpha\theta^{(1)}}{\pi} & \frac{\alpha\theta^{(1)}}{\pi} & -\sqrt{\frac{\mu_0\varepsilon^{(1)}}{\varepsilon_0\mu^{(1)}}} & -\sqrt{\frac{\mu_0\varepsilon^{(1)}}{\varepsilon_0\mu^{(1)}}} \\ 0 & 0 & -\frac{k_z^{(1)}}{\varepsilon^{(1)}}\sqrt{\frac{\varepsilon^{(1)}}{\mu^{(1)}}} & \frac{k_z^{(1)}}{\varepsilon^{(1)}}\sqrt{\frac{\varepsilon^{(1)}}{\mu^{(1)}}} \end{pmatrix} \begin{pmatrix} E_y^{1+} \\ E_y^{1-} \\ -\mu^{(1)}c^{(1)}H_y^{1+} \\ -\mu^{(1)}c^{(1)}H_y^{1-} \end{pmatrix} =$$

$$\begin{pmatrix} 1 & 1 & 0 & 0 \\ -\frac{k_z^{(2)}}{\mu^{(2)}} & \frac{k_z^{(2)}}{\mu^{(2)}} & -\frac{\alpha\theta^{(2)}}{\pi}\sqrt{\frac{\mu_0\varepsilon^{(2)}}{\varepsilon_0\mu^{(2)}}}\frac{k_z^{(2)}}{\varepsilon^{(2)}} & \frac{\alpha\theta^{(2)}}{\pi}\sqrt{\frac{\mu_0\varepsilon^{(2)}}{\varepsilon_0\mu^{(2)}}}\frac{k_z^{(2)}}{\varepsilon^{(2)}} \\ \frac{\alpha\theta^{(2)}}{\pi} & \frac{\alpha\theta^{(2)}}{\pi} & -\sqrt{\frac{\mu_0\varepsilon^{(2)}}{\varepsilon_0\mu^{(2)}}} & -\sqrt{\frac{\mu_0\varepsilon^{(2)}}{\varepsilon_0\mu^{(2)}}} \\ 0 & 0 & -\frac{k_z^{(2)}}{\varepsilon^{(2)}}\sqrt{\frac{\varepsilon^{(2)}}{\mu^{(2)}}} & \frac{k_z^{(2)}}{\varepsilon^{(2)}}\sqrt{\frac{\varepsilon^{(2)}}{\mu^{(2)}}} \end{pmatrix} \begin{pmatrix} E_y^{2+} \\ E_y^{2-} \\ -\mu^{(2)}c^{(2)}H_y^{2+} \\ -\mu^{(2)}c^{(2)}H_y^{2-} \end{pmatrix}. \tag{3}$$

Therefore, the wave function vectors $\Xi^{(1)} = \left(E_y^{1+}, E_y^{1-}, -\mu^{(1)}c^{(1)}H_y^{1+}, -\mu^{(1)}c^{(1)}H_y^{1-}\right)^T$, $\Xi^{(2)} = \left(E_y^{2+}, E_y^{2-}, -\mu^{(2)}c^{(2)}H_y^{2+}, -\mu^{(2)}c^{(2)}H_y^{2-}\right)^T$ are related by $\Xi^{(1)} = \boldsymbol{M}^{(21)}\Xi^{(1)}$, where the elements of the transfer matrix $\boldsymbol{M}^{(21)}$ are given by

$$M_{11}^{(21)} = M_{22}^{(21)} = \frac{1}{2}\left(1 + z_{(s)}^{(21)}\right),\; M_{12}^{(21)} = M_{21}^{(21)} = \frac{1}{2}\left(1 - z_{(s)}^{(21)}\right), \tag{4a}$$

$$M_{13}^{(21)} = M_{24}^{(21)} = -M_{23}^{(21)} = -M_{14}^{(21)} = \frac{1}{2}\sqrt{\frac{\mu_0\varepsilon^{(1)}}{\varepsilon_0\mu^{(1)}}}\, z_{(s)}^{(21)}\delta_{12}, \tag{4b}$$

$$M_{33}^{(21)} = M_{44}^{(21)} = \frac{1}{2}\sqrt{\frac{\varepsilon^{(1)}\mu^{(2)}}{\varepsilon^{(2)}\mu^{(1)}}}\left(1 + z_{(p)}^{(21)}\right),\; M_{34}^{(21)} = M_{43}^{(21)} = \frac{1}{2}\sqrt{\frac{\varepsilon^{(1)}\mu^{(2)}}{\varepsilon^{(2)}\mu^{(1)}}}\left(1 - z_{(p)}^{(21)}\right), \tag{4c}$$

$$M_{31}^{(21)} = M_{32}^{(21)} = M_{41}^{(21)} = M_{42}^{(21)} = -\frac{1}{2}\sqrt{\frac{\varepsilon_0\mu^{(2)}}{\mu_0\varepsilon^{(2)}}}\delta_{12}, \tag{4d}$$

$$z_{(s)}^{(21)} = \frac{\mu^{(2)}k_z^{(1)}}{\mu^{(1)}k_z^{(2)}},\; z_{(p)}^{(21)} = \frac{\varepsilon^{(2)}k_z^{(1)}}{\varepsilon^{(1)}k_z^{(2)}},\; \delta_{12} = \frac{\alpha\left(\theta^{(1)} - \theta^{(2)}\right)}{\pi}. \tag{4e}$$

Note that $\frac{1}{2}\left(\left(\left(1 + z_{(s)}^{(21)}\right),\left(1 - z_{(s)}^{(21)}\right)\right)^T, \left(\left(1 - z_{(s)}^{(21)}\right),\left(1 + z_{(s)}^{(21)}\right)\right)^T\right)$ is the transfer matrix of pure transversal electric (TE), whereas $\frac{1}{2}\left(\left(\left(1 + z_{(p)}^{(21)}\right),\left(1 - z_{(p)}^{(21)}\right)\right)^T, \left(\left(1 - z_{(p)}^{(21)}\right),\left(1 + z_{(p)}^{(21)}\right)\right)^T\right)$ is the transfer matrix of pure transversal magnetic (TM) modes arising in the absence of topological term [15]. Note also that the terms associated to the topological parameters $\theta^{(1)}$, $\theta^{(2)}$, couple the TE and TM modes. Such terms depend on the jump $\theta^{(1)} - \theta^{(2)}$ of the topological parameter at the considered surface and they are linear with respect to the fine structure constant.

3. Results and discussion

General relations (4a-c) can be used for the description of several electrodynamic properties of systems containing TI media. Let us consider the case when the region $z<0$ of the space is vacuum, with $\varepsilon^{(1)} = \varepsilon_0$, $\mu^{(1)} = \mu_0$, and $\theta^{(1)} = 1$, whereas the region $z>0$ is occupied by a TI material with $\mu^{(2)} = \mu_0$, $\theta^{(2)} = \pi$ and refraction index $n^{(2)} = \sqrt{\varepsilon^{(2)}/\varepsilon_0}$.

Suppose a TE mode electromagnetic wave with amplitude E_y^{1+}, which is incident from the region $z<0$ with an incident angle ϑ_1; in the region $z>0$ we have only forward fields. Then

$\Xi^{(1)}=\left(E_y^{1+},E_y^{1-},0,-\mu^{(1)}c^{(1)}H_y^{1-}\right)^T$, $\Xi^{(2)}=\left(E_y^{2+},0,-\mu^{(2)}c^{(2)}H_y^{2+},0\right)^T$ and the relation $\Xi^{(1)}=\boldsymbol{M}^{(21)}\Xi^{(1)}$ leads to the following expressions for the amplitudes of the reflected and transmitted TE and TM modes:

$$E_y^{1-}=r_{ss}^{(21)}E_y^{1+},\ E_y^{2+}=\left[M_{11}^{(21)}+M_{12}^{(21)}r_{ss}^{(21)}-M_{14}^{(21)}r_{sp}^{(21)}\right]E_y^{1+}, \tag{5a}$$

$$H_y^{1-}=\sqrt{\frac{\varepsilon^{(1)}}{\mu^{(1)}}}r_{sp}^{(21)}E_y^{1+},\ H_y^{2+}=\sqrt{\frac{\varepsilon^{(2)}}{\mu^{(2)}}}\left[-M_{31}^{(21)}-M_{32}^{(21)}r_{ss}^{(21)}+M_{34}^{(21)}r_{sp}^{(21)}\right]E_y^{1+}, \tag{5b}$$

$$r_{ss}^{(21)}=\frac{M_{21}^{(21)}M_{44}^{(21)}-M_{24}^{(21)}M_{41}^{(21)}}{M_{22}^{(21)}M_{44}^{(21)}-M_{42}^{(21)}M_{24}^{(21)}},\ r_{sp}^{(21)}=\frac{M_{21}^{(21)}M_{42}^{(21)}-M_{22}^{(21)}M_{41}^{(21)}}{M_{22}^{(21)}M_{44}^{(21)}-M_{42}^{(21)}M_{24}^{(21)}}. \tag{5c}$$

In the expressions (5c) $r_{ss}^{(21)}$ ($r_{sp}^{(21)}$) is the reflection amplitude of an incident TE mode which is reflected in a TE (TM) mode.

In a similar way, for a TM mode electromagnetic wave with amplitude H_y^{1+} the amplitudes of the reflected and transmitted TM and TE modes are given by:

$$H_y^{1-}=r_{pp}^{(21)}H_y^{1+},\ H_y^{2+}=\sqrt{\frac{\mu^{(1)}\varepsilon^{(2)}}{\varepsilon^{(1)}\mu^{(2)}}}\left(-M_{32}^{(21)}r_{sp}^{(21)}+M_{33}^{(21)}+M_{34}^{(21)}r_{pp}^{(21)}\right)H_y^{1+}, \tag{6a}$$

$$E_y^{1-}=\sqrt{\frac{\mu^{(1)}}{\varepsilon^{(1)}}}r_{sp}^{(21)}H_y^{1+},\ E_y^{2+}=\sqrt{\frac{\mu^{(1)}}{\varepsilon^{(1)}}}\left[M_{12}^{(21)}r_{sp}^{(21)}-M_{13}^{(21)}-M_{14}^{(21)}r_{pp}^{(21)}\right]H_y^{1+}, \tag{6b}$$

$$r_{pp}^{(21)}=\frac{M_{23}^{(21)}M_{42}^{(21)}-M_{22}^{(21)}M_{43}^{(21)}}{M_{22}^{(21)}M_{44}^{(21)}-M_{42}^{(21)}M_{24}^{(21)}},\ r_{sp}^{(21)}=\frac{M_{43}^{(21)}+M_{44}^{(21)}}{M_{22}^{(21)}M_{44}^{(21)}-M_{42}^{(21)}M_{24}^{(21)}}M_{23}^{(21)}. \tag{6c}$$

In (6c) $r_{pp}^{(21)}$ ($r_{ps}^{(21)}$) is the reflection amplitude of an incident TM mode which is reflected in a TM (TE) mode.

Note that $\boldsymbol{R}^{(21)}=\left(\left(r_{ss}^{(21)},r_{ps}^{(21)}\right)^T,\left(r_{sp}^{(21)},r_{pp}^{(21)}\right)^T\right)$ is the reflection matrix describing the relation between the incident and the reflection fields at the single interface of TI from the medium 1 to medium 2 [9, 10].

By taking into account the explicit relations (4a-d) for the elements of the transfer matrix, the reflection coefficients of the considered modes are

$$R_{pp}^{(21)}(\delta_{12})=\left|r_{pp}^{(21)}\right|^2\approx R_p\left[1+\left(\frac{4z_{(s)}^{(21)}z_{(p)}^{(21)}}{\left(1+z_{(s)}^{(21)}\right)\left(1+z_{(p)}^{(21)}\right)^2}\right)\delta_{12}^2\right], \tag{7a}$$

$$R_{ss}(\delta_{12})=\left|r_{ss}^{(21)}\right|^2\approx R_s\left[1+\left(\frac{4z_{(s)}^{(21)2}}{\left(1+z_{(p)}^{(21)}\right)\left(1+z_{(s)}^{(21)}\right)^2}\right)\delta_{12}^2\right], \tag{7b}$$

$$R_{ps}^{(21)}=R_{sp}^{(21)}=\left|r_{ps}^{(21)}\right|^2\approx\frac{4z_{(s)}^{(21)2}}{\left(1+z_{(s)}^{(21)}\right)^2\left(1+z_{(p)}^{(21)}\right)^2}\delta_{12}^2, \tag{7c}$$

where $R_{s,p}=\left|\frac{\left(1-z_{(s,p)}^{(21)}\right)}{\left(1+z_{(s,p)}^{(21)}\right)}\right|^2$ are the reflection coefficients of TE and TM waves in conventional dielectrics [17].

It can be seen that the changes in reflection coefficients $\delta R_{pp}^{(21)}$, $\delta R_{ss}^{(21)}$ which are induced by the non trivial topology vary as quadratic functions of the topological term. Although these corrections are small, they show some special features, which are evident from Fig. 1:

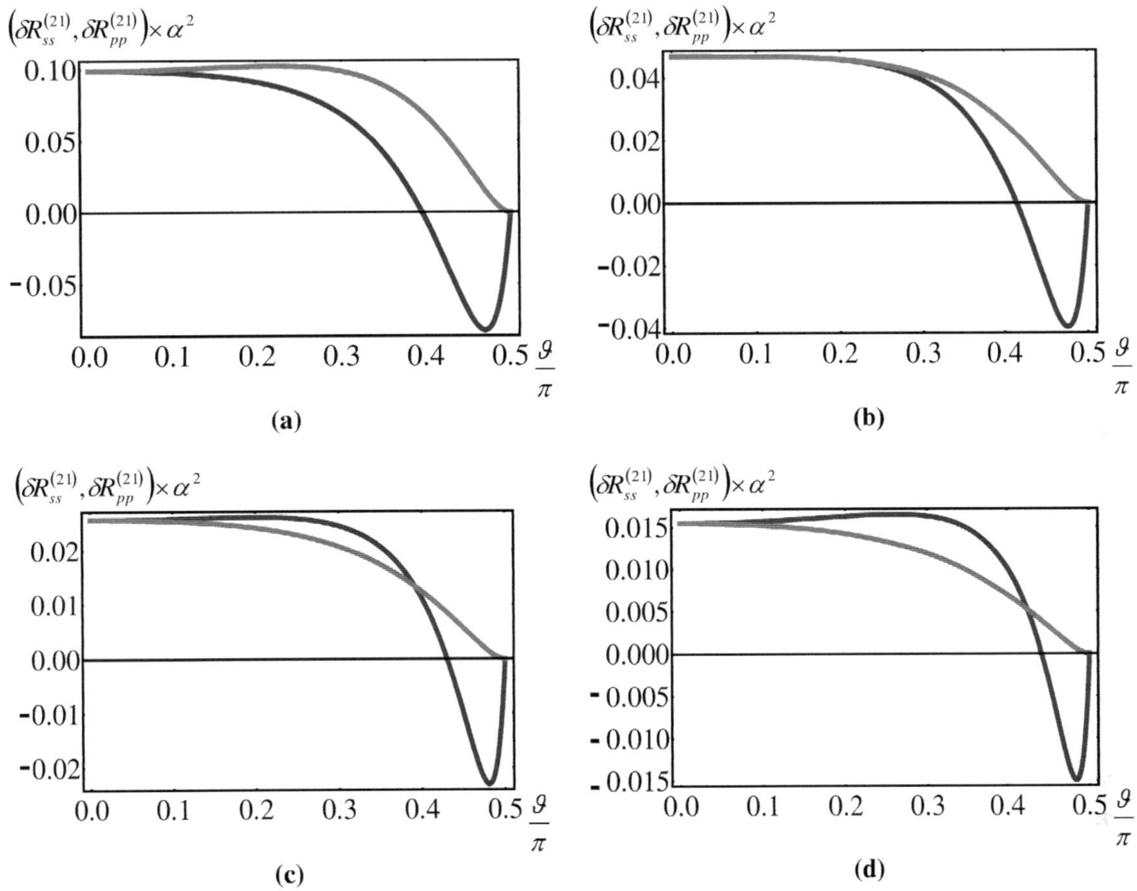

Figure 1. Change (associated to the presence of a jump in the axion term) in the reflection coefficient of TE (red curves, $\delta R_{ss}^{(21)}$) and TM (blue curves, $\delta R_{pp}^{(21)}$) waves incident from the region z<0 as a function of the incident angle ϑ_1. The region z<0 of the space is vacuum, with $\varepsilon^{(1)} = \varepsilon_0$, $\mu^{(1)} = \mu_0$, and $\theta^{(1)} = 1$, whereas the region z<0 is occupied by a TI material with $\mu^{(2)} = \mu_0$, $\theta^{(2)} = \pi$ and refraction index $n^{(2)}$ *(a)* 2, *(b)* 3, *(c)* 4 and *(d)* 5.

(i) The magnitude of the change $\delta R_{ss}^{(21)}$ of the reflection coefficient for TE modes is a slowly varying function of the incident angle ϑ_1 and it vanishes only at the incidence angle $\vartheta_1 = \pi/2$.

(ii) For TM modes the change $\delta R_{pp}^{(21)}$ vanishes additionally at an incidence angle $\tilde{\vartheta} < \pi/2$ and it has an extremum at $\tilde{\vartheta} < \vartheta_1 < \pi/2$. The positions of both the zero and the extremum of $\delta R_{pp}^{(21)}$ shifts toward $\pi/2$ as well as the refraction index of the TI media increases.

(iii) As well as the refraction index contrast increases, the correction $\delta R_{pp}^{(21)}$ becomes greater than $\delta R_{ss}^{(21)}$ in a wide interval of incidence angles.

It is worthy to mention that $R_{pp}^{(21)}(\delta_{12})$ given by (7a) vanishes at the incidence angle given by

$$\tilde{\vartheta}_B = \tan^{-1}\left[\left(1 - \frac{2n^{(1)2}n^{(2)2}}{\left(n^{(1)2}+n^{(2)2}\right)^2}\delta_{12}^2\right)\left(\frac{n^{(1)}}{n^{(2)}}\right)\right], \tag{8}$$

which reduces to the Brewster angle $\vartheta_B = \tan^{-1}\left(\frac{n^{(1)}}{n^{(2)}}\right)$ in the absence of jump in the axion term. The transmission angle $\vartheta^{(2)} = \sin^{-1}\left(\frac{n^{(1)}}{n^{(2)}}\sin\tilde{\vartheta}_B\right)$ is, therefore,

$$\vartheta^{(2)} = \frac{\pi}{2} - \tilde{\vartheta}_B + \frac{n^{(2)}}{n^{(1)}}\frac{2n^{(1)2}n^{(2)2}}{\left(n^{(1)2}+n^{(2)2}\right)^2}\delta_{12}^2. \tag{9}$$

This result indicates that for the incidence of a TM wave with incidence angle $\tilde{\theta}_B$, the propagation vector of the reflected wave is no longer perpendicular to that of the transmitted wave, which is in agreement with [8].

The expressions for the 4×4 transfer matrix (4a-d) can be used to obtain the nontrivial Kerr (Faraday) rotation angle ν_K (ν_F) of light reflected (transmitted) off an interface with jump in the topological axion parameter. For an incident TM wave these rotation angles are given by

$$\tan(\nu_K) = \frac{E_y^{1-}}{E_x^{1-}} = -\frac{\omega}{c^{(1)}k_z^{(1)}}\frac{r_{sp}}{r_{pp}}, \quad \tan(\nu_F) = \frac{E_y^{2+}}{E_x^{2+}} = \frac{\omega}{c^{(2)}k_z^{(2)}}\frac{\left(M_{12}r_{sp} - M_{13} - M_{14}r_{pp}\right)}{\left(-M_{32}r_{sp} + M_{33} + M_{34}r_{pp}\right)}. \tag{10}$$

Note that these magnitudes are linear functions of the parameter $\delta_{12} = \frac{\alpha\left(\theta^{(1)} - \theta^{(2)}\right)}{\pi}$ [12, 18], but their behavior depend additionally on the relation between the incident $\vartheta^{(1)}$ and the Brewster $\tilde{\vartheta}_B$ angles, as is illustrated in Fig. 2 for the parameter relations of Fig. 1b. Note that the magnitude of ν_K increases for $\vartheta^{(1)} < \tilde{\vartheta}_B$ whereas it decreases for $\tilde{\vartheta}_B < \vartheta^{(1)} < \pi/2$ and vanishes at $\vartheta^{(1)} = \pi/2$; additionally, the direction of rotation of the polarization plane of the reflected light changes when the angle of the incident TM wave passes through the Brewster angle. On the other hand, the direction of rotation of the polarization plane of the transmitted light is conserved in the whole interval $0 < \vartheta^{(1)} < \pi/2$, but ν_F changes as a non-monotonous function of the incident angle $\vartheta^{(1)}$ and vanishes at $\vartheta^{(1)} = \pi/2$.

IOP Conf. Series: Journal of Physics: Conf. Series **850** (2017) 012024 doi:10.1088/1742-6596/850/1/012024

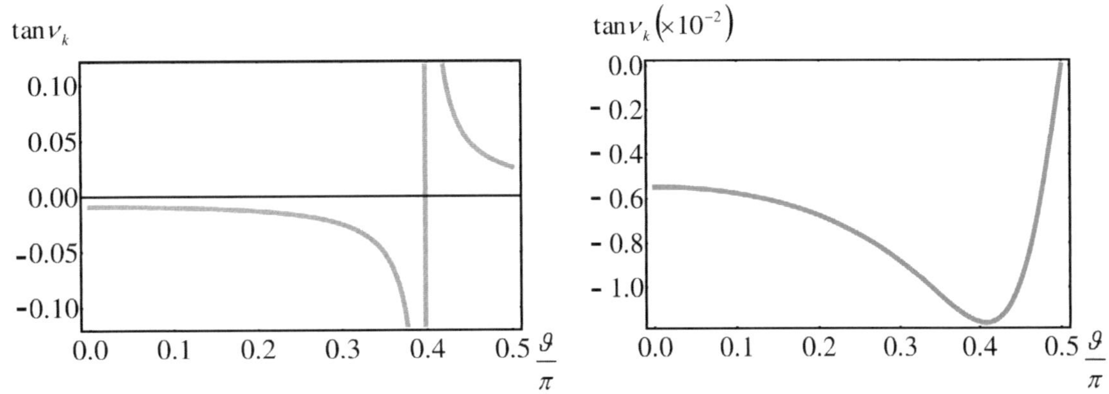

(a) **(b)**

Figure 2. Behavior (as functions the incident $\vartheta^{(i)}$ angle of an incident TM wave) of the angles of rotation of polarization planes of reflected *(a)* and transmitted *(b)* waves, for the parameter relations of Fig. 1b.

4. Conclusion

Some electrodynamic properties of systems containing topological insulators were described on the basis of the analytical expressions obtained in the frame of axion electrodynamics for the components of the transfer matrix for the interface between a conventional dielectric and a topological insulator. TE and TM modes are coupled by matrix terms which are linear with respect to the jump of the axion topological parameter. Analytical expressions were obtained for the changes induced by the non trivial topology in the reflection coefficients from TM and TE incident waves and for the rotation of the polarization planes of reflected and transmitted waves.

The obtained relations can be applied for the description of more complex geometries, by extending the well-known techniques developed for conventional dielectrics [15, 16].

Acknowledgements
This work is part of the research project 1106-712-49884 (contract No.264-2016), supported by COLCIENCIAS (FONDO NACIONAL DE FINANCIAMIENTO PARA LA CIENCIA. LA TECNOLOGÍA Y LA INNOVACIÓN "FRANCISCO JOSÉ DE CALDAS"). D F Rojas is supported by COLCIENCIAS by the program JOVENES INVESTIGADORES. JCG thanks the Excellence Center of New Materials (CENM) for financial support of the research group.

References
[1] Hasan M Z and Kane C L 2010 *Rev. Mod. Phys.* 82 3045
[2] Qi X L and Zhang S C 2011 *Rev. Mod. Phys.* 83 1057
[3] Shen S Q 2012 *Topological Insulators: Dirac Equation in Condensed Matters* (Springer-Verlag Berlin Heidelberg)
[4] Franz M and Molenkamp L 2013 *Topological Insulators* (Elsevier, Oxford B.V.)
[5] Bernevig B A and Hughes T L 2013 *Topological Insulators and Topological Superconductors* (Princeton University Press, Princeton)
[6] Asbóth J K, Oroszlány L and Pályi A 2016, *A Short Course on Topological Insulators*, (Springer, Heidelberg)
[7] Qi X L, Hughes T L and Zhang S-C 2008 *Phys. Rev.* B **78** 195424

[8] Chang M C and Yang M F 2009 *Phys. Rev.* B **80** 113304
[9] Nie W, Zeng R, Lan Y and Zhu S 2013 *Phys. Rev.* B **88** 085421
[10] Zeng R, Chen L, Nie W, Bi M, Yang Y and S Zhu 2016 *Phys. Lett.* A **380** 2861
[11] Karch A 2009 *Phys. Rev. Lett.* **103** 171601
[12] Karch A 2011 *Phys. Rev.* B. **83** 245432
[13] Granada J C, Rojas D F 2014 *Physica* B **455** 82
[14] Lai Y P, Lin I T, Wu K H and Liu J M 2014 *Nanomaterials and Nanotechnology* **4** 13
[15] Markoš P.and Soukoulis C. M. 2008 *Wave Propagation. From Electrons to Photonic Crystals and Left-Handed Materials* Princeton university press (Oxford)
[16] Sánchez-Soto L L, Monzón J ., Barriuso A G and Cariñena JF 2012 *Physics Reports* **513** 191
[17] Born M and Wolf E 2003 *Principles of Optics: Electromagnetic Theory of Propagation, Interference and Diffraction of Light*, seventh (expanded) ed. Cambridge Univ. Press, Cambridge

Use of optical vortices for the compression of fingerprint recognition systems

Martín Rodríguez-Vega and René Restrepo

Applied Optics Group, Physical Sciences Department, School of Science, Universidad EAFIT, Medellín, Colombia

E-mail: mrodri39@eafit.edu.co, rrestre6@eafit.edu.co

Abstract. In this article, an algorithm is presented for the processing and verification of fingerprints by analyzing optical vortices present in the orientation of the fringes of the fingerprint. The algorithm presents a compression factor of 1200, but is highly susceptible to the presence of noise.

1. Introduction

It is well known that fingerprint verification is one of the most popular and broadly used biometric technologies. Because of its success, there exists large databases that contain the information of millions of users [1]. Therefore, methods and algorithms that yield high compression factors whilst allowing the reliable recognition of a person's identity are required [2]. Our work was motivated by the work of Larkin and Fletcher [2], who obtained useful information from the amplitude and phase of fingerprints using an holographic analysis framework. Minutiae location was shown to be related with phase singularities. After a revision of their work, we propose a similar yet different approach to reduce the information that needs to be stored, while reducing the number of calculations. We also detected problems in some of the steps required for phase demodulation, and we propose a way to omit this process.

2. Background

Larkin and Fletcher [2] developed a fingerprint analysis framework based on interferometry where fingerprint images are represented as holograms: $f(x,y) = a(x,y) + b(x,y)\cos(\psi(x,y))$. One of the common methods used for fingerprint classification and recognition consist on locating the minutiae of the fingerprint [3, 4]. Minutiae are characteristic details such as bifurcations, abrupt ends and loops. From interferometry, it is known that this details can be modeled as phase singularities, also called optical vortices, in the phase of the fringe pattern [2, 5].

The fringe pattern representation of the fingerprint image can be understood as the in-phase component of a complex optical field. Thus, for obtaining the desired phase, the quadrature component of the field is required. Larkin [6, 7], proposed an isotropic two-dimensional radial extension of the Hilbert transform for fringe patterns. The proposed transform, named the vortex operator, consists of multiplying the signal with a spiral phase kernel in the Fourier domain ($\mathcal{V}\{g(x,y)\} = \mathcal{F}^{-1}\{e^{i\theta(u,v)}\mathcal{F}\{g(x,y)\}\}$).

For a fringe pattern, Larkin [6] showed that the vortex operator obtained the quadrature of the field with an additional complex term, $\mathcal{V}\{b(x,y)\cos(\psi(x,y))\} \approx ie^{i\beta(x,y)}b(x,y)\sin(\psi(x,y)$.

The appearance of the direction field $\beta(x, y)$ is expected as direction estimation has been found to be important for fringe pattern and fingerprint processing [5]. However, this calculation isn't trivial: the direction of a local point in a fringe pattern cannot be usually differentiated from a π radian shifted version of itself, due to the even behavior of the cosine function. Thus, the direction map modulo 2π cannot be determined and only the orientation map modulo π can be estimated.

Larkin [8] proposed a method for orientation estimation through the use of a modified version of the vortex operator. This new operator, named the energy operator, was shown to be more robust to the presence of noise than other traditional gradient based methods, $\left(i\mathcal{V}\{g(x,y)\}\right)^2 + g(x,y)\mathcal{V}\{\mathcal{V}\{g(x,y)\}\} = b(x,y)^2 e^{2i\beta(x,y)}$. Although the direction map cannot be generally known, when the phase map is smooth and there are no singularities, some methods can be applied to recover the direction from the orientation [7, 5, 9].

Once obtained the direction map, the phase of the fringe pattern can be calculated from the in-phase and quadrature components. As discussed earlier, the phase contains singularities which are associated to minutiae in the fingerprint pattern. The detection of these singularities can be achieved by the residue theorem for phase unwrapping [9, 10], $\oint_C \nabla\psi(x,y) \cdot d\vec{l} = 2\pi \sum_{k=1}^{N} p_k$, where N is the number of singularities inside a close circuit C and $p_k \in \mathcal{Z}$ is the charge of the kth singularity. For fingerprints, the charge of the singularities are restricted to ± 1 as discussed in [2].

3. Use of phase for locating singularities

When the method described in the previous section was implemented, we found that direction estimation from the orientation map had a significant impact on the quality of the demodulated phase. Using the energy operator proposed in [8] and described in the previous section on a sample fingerprint image, the wrapped orientation map in Fig. (1a) was obtained. For direction recovery, the orientation was first unwrapped as shown in Fig. (1b). It can be seen that new structures appeared in the image: the horizontal streaks that go from within the fingerprint to the edges indicate the presence of vortices. It is known that phase singularities make the unwrapping process difficult, as the result becomes path dependent [9, 10]. The resulting phase map is shown in Fig. (1c) which presents streaks that induce a sign shift between the upper and the lower values. These streaks cause problems to locate the singularities in the phase map: false vortices are detected in the intersection between streaks and fringes.

There are some possible solutions to remove the sign shift caused by these singularities. In [10], a Helmholtz decomposition of the phase map in the "smooth" and "singular" components is proposed. However, as the singularities are detected using 2x2 pixel integration loops, there is uncertainty in the exact location of the vortex, which greatly increases the difficulty of the Helmholtz decomposition. Another solution is to use branch cuts [10] to minimize the effect of unbalanced singularities when the orientation is unwrapped. If the branch cuts could be oriented in such a way that the sign inverting streaks were parallel to the fringes, then the presence of false vortices would be avoided. Nevertheless, this would greatly increase the computational cost of phase unwrapping, and in some cases (e.g. the loop structure in the center of Fig. (1c)) it is impossible to design branch cuts that don't intersect with fingerprint fringes. An alternative approach is suggested after a number of observations of orientation and phase maps. The motivation for this approach is based upon the following question: **is there any relationship between the singularities present in the orientation map and those present in the phase map?**

4. Use of orientation for locating singularities

Once we evidenced the underlying problems of the fringe pattern's direction estimation, we analyzed the physical meaning of the singularities found in the orientation map. In [5] it is

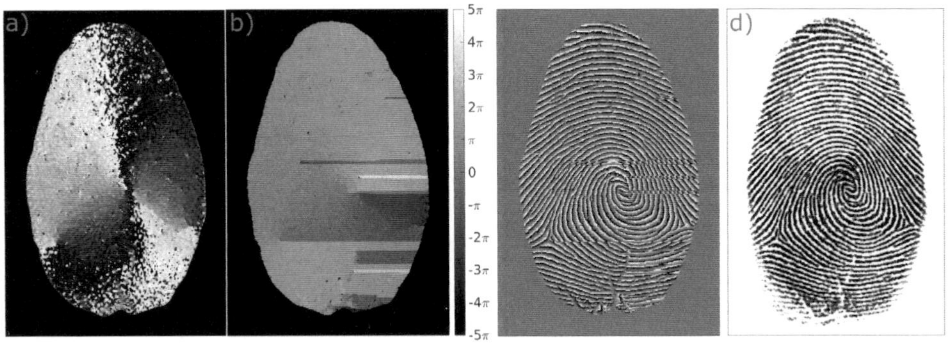

Figure 1: (a) Wrapped orientation map. (b) Unwrapped orientation map. (c) Demodulated phase from the unwrapped orientation. (d) Singularities found in the orientation map.

shown that certain fingerprint characteristics, such as deltas and loops, cause singularities in the local fringe orientation. As the phase map and its orientation are logically related, can the vortices in the orientation be assigned to fingerprint minutiae?

Vortex identification was performed over the orientation map, as shown in Fig. (1d). It can be seen that the location of the singularities (blue asterisks in the figure) in fact correspond to minutiae location in the fingerprint. Thus, knowledge of orientation suffices for minutiae identification. This results in a considerable reduction in complexity than the method presented by Larkin and Fletcher [2]. However, it was observed that only the positive charges of the singularities correspond to fingerprint minutiae. Therefore, although we no longer require to calculate the phase of the fringe pattern, we loose the information of the polarity of the singularities.

5. Proposed algorithm

In this section, we present an algorithm for minutiae detection for fingerprint recognition. Although the algorithm's structure is similar to the one proposed in [2], our approach has the objective of further compressing the information for recognition purposes and simplify the required calculations, but disregarding the reconstruction stage.

(i) **Step 1:** This step performs a first quality-enhancement preprocessing of the fingerprint image. This is required to remove the offset $(a(x, y))$. For our analysis, a simple bandpass filter proved to give good results, but more complex methods can be used [11].

(ii) **Step 2:** This step will separate the amplitude modulation and phase modulation components. We use the vortex operator on the filtered signal and take the magnitude of the output. This yields $b(x, y)|\sin(\psi(x, y))|$. If we take the arctangent of this signal with the output of **Step 1** we obtain an estimation of the phase $\psi(x, y)$ with a sign ambiguity due to the absolute value of the quadrature. However, as cosine is an even function, we can recover the in-phase component of the fringe pattern by taking the cosine of the estimation.

(iii) **Step 3:** In this step, we use the energy operator on the output of **Step 2**. As the magnitude of the input is 1, the output is just the complex signal related to the orientation $\exp(i2\beta(x, y))$. Taking the arctangent of the complex field, the orientation phase map $2\beta(x, y)$ is obtained.

(iv) **Step 4:** This step will locate the phase singularities present in the orientation map. For this, the residue theorem is used. As the orientation is assumed to be discretized in a matrix, the smallest possible closed circuit is a 2x2 pixel loop. For each position (m, n) in a matrix M, the presence of a singularity is given by: $\sum_{C_{2x2}} \nabla(2\beta(x, y)) \cdot \vec{dl} = p_k$.

(v) **Step 5:** Identify and save the (m, n) coordinates of the positive singularities. The relative positions of the vortices can be then used for fingerprint recognition [3, 4].

6. Discussion

In summary, the previous sections developed a method for locating singularities (associated with minutiae) in the orientation map of a fingerprint image. The input of the proposed algorithm is the scanned fingerprint image, and the output is the location of the positive singular values of the orientation map. Because of this, the information required to be stored for identification purposes is greatly compressed. The compression factor will depend on the number of singularities present in the orientation map and the maximum size of the image. The maximum compression factor we obtained was of 1200 for a 1200x800 pixels image, whilst the mean compression factor was approximately 500. However, there are some limitations and trade-offs that need to be taken into account. As discussed in the previous section, this work does not take into consideration the reconstruction of fingerprints, as only the minimum information possible is stored. This limits the possible uses but it is considered that applications that deal only on person verification and that are subject to heavily limited storage capacity could benefit from this method. Additionally, it was observed that noise and image quality had a very powerful effect on the results of the algorithm. Noisy images that present dirty or diffuse areas create false ridges that generate a high number of singularities. Also, the presence of pores is in some cases classified by the algorithm as short bifurcations and junctions of fringes, thus allocating a pair of singularities. It is suggested that this method is used along with high resolution fingerprint scanners: as the compression factor is high, this scanners can be used without the need to store a high amount of information.

7. Conclusion

This work examined the problems underlying the process of phase demodulation for locating the phase singularities associated with fingerprint minutiae. The presence of vortices in the fringe orientation induced a problematic phase unwrapping, which was observed with sign shifting streaks in the demodulated phase, that in turn originated false singularities. After analysis of the singularities in the orientation map, it was observed that these vortices corresponded as well with minutiae. It is proposed to calculate the vortex locations from the orientation map instead of the phase map. The resulting method and algorithm yield very high compression factors, but also evidenced to be greatly affected by noise and image quality.

References

[1] Sankaran A, Agarwal A, Keshari R, Ghosh S, Sharma A, Vatsa M and Singh R 2015 *2015 IEEE 7th International Conference on Biometrics Theory, Applications and Systems, BTAS 2015* 1–6
[2] Larkin K G and Fletcher P A 2007 *Optics express* **15** 8667–77 ISSN 1094-4087
[3] Uz T, Bebis G, Erol A and Prabhakar S 2009 *Computer Vision and Image Understanding* **113** 979–992 ISSN 10773142
[4] Zhang Q, Yin Y and Yang G 2016 *Neurocomputing* **171** 1401–1413 ISSN 18728286
[5] Servin M, Quiroga J A and Padilla J M 2014 *Fringe Pattern Analysis for Optical Metrology* (Weinheim, Germany: Wiley-VCH) ISBN 978-3-527-41152-8
[6] Larkin K G, Bone D J and Oldfield M A 2001 *Journal of the Optical Society of America A* **18** 1862 ISSN 1084-7529
[7] Larkin K G 2001 *Journal of the Optical Society of America A* **18** 1871 ISSN 1084-7529
[8] Larkin K 2005 *Optics express* **13** 8097–8121 ISSN 1094-4087
[9] Ying L 2006 *Wiley Encyclopedia of Biomedical Engineering* vol 24 (Hoboken, NJ, USA: John Wiley & Sons, Inc.) pp 1–11 ISBN 0471740365
[10] Ghiglia D C and Pritt M D 1998 *Two-Dimensional Phase Unwrapping: Theory, Algorithms, and Software* (Wiley-Interscience) ISBN 978-0-471-24935-1
[11] Willis A and Myers L 2001 *Pattern Recognition* **34** 255–270 ISSN 00313203

AUTHOR INDEX